Remaking Regional Economi

Since the early 1980s, the region has been central to thinking about the emerging character of the global economy. In fields as diverse as business management, industrial relations, economic geography, sociology, and planning, the regional scale has emerged as an organizing concept for interpretations of economic change.

This book is both a critique of the "new regionalism" and a return to the "regional question," including all of its concerns with equity and uneven development. It will challenge researchers and students to consider the region as a central scale of action in the global economy. At the core of the book are case studies of two industries that rely on skilled, innovative, and flexible workers – the optics and imaging industry and the film and television industry. Combined with this is a discussion of the regions that constitute their production centers. The authors' intensive research on photonics and entertainment media firms, both large and small, led them to question some basic assumptions behind the new regionalism and to develop an alternative framework for understanding regional economic development policy. Finally, there is a re-examination of what the regional question means for the concept of the learning region.

This book draws on the rich contemporary literature on the region but also addresses theoretical questions that preceded "the new regionalism." It will contribute to teaching and research in a range of social science disciplines.

Susan Christopherson is a professor at the Department of City and Regional Planning, Cornell University.

Jennifer Clark is an assistant professor at the School of Public Policy, Georgia Institute of Technology.

Routledge Studies in Economic Geography

The Routledge Studies in Economic Geography series provides a broadly based platform for innovative scholarship of the highest quality in economic geography. Rather than emphasizing any particular sub-field of economic geography, we seek to publish work across the breadth of the field and from a variety of theoretical and methodological perspectives.

Published:
Economic Geography: Past, present and future
Edited by Sharmistha Bagchi-Sen and Helen Lawton Smith

Remaking Regional Economies: Power, labor, and firm strategies in the knowledge economy
Susan Christopherson and Jennifer Clark

Forthcoming:

The New Economy of the Inner City: Regeneration and dislocation in the twenty first century metropolis
Thomas Hutton

Remaking Regional Economies

Power, labor, and firm strategies
in the knowledge economy

By Susan Christopherson
and Jennifer Clark

LONDON AND NEW YORK

First published 2007
by Routledge
2 Park Square, Milton Park, Abingdon, Oxfordshire OX14 4RN

Simultaneously published in the USA and Canada
by Routledge
711 Third Avenue, New York, NY 10017

First issued in paperback 2016

Routledge is an imprint of the Taylor & Francis Group, an informa business

Typeset in Galliard
by Keystroke, 28 High Street, Tettenhall, Wolverhampton

British Library Cataloguing in Publication Data
A catalogue record for this book is available from the British Library

Library of Congress Cataloging in Publication Data
Christopherson, Susan.
Remaking regional economies : power, labor, and firm strategies in the knowledge economy / by Susan Christopherson and Jennifer Clark. – 1st ed.
p. cm.
1. Regional economies. 2. Regional planning. 3. Technological innovations–Economic aspects.
I. Clark, Jennifer, 1972– II. Title.
HT388.C483 2007
338'.064–dc22
2007011476

ISBN 13: 978-1-138-98507-0 (pbk)
ISBN 13: 978-0-415-35743-2 (hbk)

Contents

Illustrations

Figures

Tables

Preface

Were it part of our everyday education and comment that the corporation is an instrument for the exercise of power, that it belongs to the process by which we are governed, there would then be debate on how that power is used and how it might be made subordinate to the public will and need. This debate is avoided by propagating the myth that the power does not exist. It is especially useful that the young be so instructed. By pretending that power is not present, we greatly reduce the need to worry about its exercise.

(Galbraith 1977: 259)

This book reflects many years of researching, teaching, and thinking about how firm strategies and policy measures influence industry structure and, ultimately, the fate of regions.

The study of regional economic development in the United States attracts a small group of academics and researchers and fewer still who engage in the time consuming and expensive empirical work required to gain a nuanced understanding of regional industrial change. Our hope is that this book, and the themes of governance, power, and equity embedded in it, will reinvigorate the kind of empirically based, and theoretically rigorous policy discussions epitomized by the work of Bennett Harrison.

In many ways this book begins on the West Bank of the Mississippi River where we both began our thinking about regional economic development – informed by the research in urban and economic geography at the University of Minnesota and the practice of economic development in the state. We were both profoundly influenced by the Minnesota approach to governance and civic involvement, and the ways in which it encourages the kinds of investment in education, the environment, and infrastructure that lead to long-term sustainable economic development. This approach has become even more exceptional in a period of devolution and the roll-back of government accountability. Although we shared this experience twenty years apart, it was the Minnesota tradition of public investment and learning through applied research that informs how we approached this research project.

Our work also benefited from economic development research in diverse places across the US: in Texas, California, Arizona, Utah, and Georgia as well as in Upstate New York and New York City. These experiences taught us about the distinctiveness of the US experience and about how difficult it is to craft economic development policy without understanding national governance systems as they interact with local and regional history, politics, and power relations. Although we anticipate that our project has implications for the regional question across countries, our research has a distinctive American cast. We hope that our research on the regional question in the US context will help foster research on US market governance as one "variety of capitalism."

What has emerged from our collaboration is a project that seeks to understand the processes shaping regional economies. We see this book as rooted in past debates on the regional question and as a beginning to new ones. We are grateful to all of those who have contributed to this project – explicitly and implicitly – over the years including: Matt Drennan, Katherine Stone, Pierre Clavel, Allan Pred, Amy Glasmeier, Karen Chapple, Jamie Peck, Tod Rutherford, Joan Fitzgerald, Nancey Green Leigh, Meric Gertler, Greg Schrock, Laura Wolf-Powers, Harley Etienne, Danielle Van Jaarsveld, Julie Silva, Yael Levitte, Kristin Larsen, Ragui Assaad, Ed Goetz, Dick Bolan, Wendy Jepson, Rosemary Batt, Rachel Webber, Norma Rantisi, Deborah Leslie, Gordon Clark, Linda McDowell, Ron Martin, Neil Wrigley, Andy Pratt, Andy Pike, David Angel, Adam Tickell, Robert Kuttner, Ann Markusen, Lowell Turner, and Robert Giloth.

In the midst of debates in economic geography, urban planning, economic development, business, and industrial and labor relations about the ways to build competitive firms, cities, regions, and nation states, we find a surprising lack of discussion about questions of distribution and democracy. Our research into regionalism as a sub-national economic development project in the United States reveals that regionalism is most often a tool, not an ideology.

This "regionalism" has thus emerged as the convergence of several schools of thought. Regionalism is century long urban planning project directed at mitigating the inefficiencies of the metropolitan system with an attempt to link suburbs and cities in formal governance as well as markets. Regionalism is also a trajectory of research and theory in economic geography. It is the search for a new model by which to understand and address the persistent problems of uneven development. Finally, regionalism has been incorporated into the "new regionalism" with its rhetoric of industry clusters, learning regions, and "innovative milieu." This trajectory is increasingly co-opted by "growth interests" as the rationale for particular kinds of development strategies focused on building the competitiveness of firms.

Unlike most books about firm strategies and economic development, regional competition and transnational corporations, work organization and locational choice, this book is about power. In particular, this book is about how the political economic power of the transnational firm shapes regions and regional competition. While firms describe themselves as global actors, the region plays an important role in their calculus, not just as a production site but as a set of opportunities to reduce risk and increase profits. Contrary to much of the literature on firm networks in regions, firm decisions and action at the regional scale do not produce a "new world order" where adversarial interests reach miraculous consensus and willingly share the costs and benefits of a mutually beneficial growth. Would that it were true. This regional project often brings with it a set of rose-colored glasses through which we no longer see questions of power, distribution, representation, and agency.

As John Kenneth Galbraith pointed out twenty-five years ago, if we construct a myth in which there is no power, we need not bother considering how it is used. In this book, we look at how power exercised at a regional scale affects firms, workers, and, ultimately, regions.

Susan Christopherson
Ithaca, New York
USA

Jennifer Clark
Atlanta, Georgia
USA

Acknowledgements

We would like to thank those who supported this research and the production of this book in all its various stages of development. A special acknowledgment is due to the series editors: David Angel, Adam Tickell, and Amy Glasmeier.

We especially thank the participants in our case studies. In Rochester: the members of the Rochester Regional Photonics Cluster, Ed Murphy, and James Winston. In New York City: John Amman, Lois Gray, James Parrot, Maria Figueroa, Damone Richardson, and Ned Rightor.

We also appreciate the contributions of students in seminar and workshop courses over the last several years at both Cornell University and the Georgia Institute of Technology. In particular we appreciate the research assistance of Esther Blodau-Konick, Sudeshna Mitra, Alexa Stephens, Wyeth Friday, Christina Chan, Kevin Waskelis, Karen Westmont, and David Perkins.

We also appreciate the organizations that provided support and funding for the empirical work included here: the United States Department of Labor's Education and Training Administration's Community Audit program, the Annie E. Casey Foundation, The New York Film, Television, and Commercial Initiative, the Rockefeller Foundation, and the Industry Studies Program of the Alfred P. Sloan Foundation.

We also thank the Department of City and Regional Planning at Cornell University as well as the Cornell School of Industrial and Labor Relations both in Ithaca and New York City. We also thank Linda Johnson and the Cornell in Washington Program for support throughout this project. Susan also thanks David Soskice and the Science Center in Berlin. Finally, thanks are also due to the School of Public Policy, the Ivan Allen College, and the Enterprise Innovation Institute at the Georgia Institute of Technology.

A special acknowledgement is due to family and the friends who supported this project. Susan thanks Morgan Thomas and Ned Rightor and Jennifer thanks Benjy Flowers, Don Clark, Linda Clark, and Joanna Clark for advice, support, and patience.

The chapter on the media and film industry includes material reprinted from Geoforum, Volume 37, Issue 5, September 2006, pages 739–751. Susan Christopherson. Behind the scenes: How transnational firms are constructing a new international division of labor in media work, with permission from Elsevier.

Section I

Shaping the regional project

1 Introduction

Since the early 1980s, the region has been central to thinking about the emerging character of the global economy. In fields as diverse as business management, industrial relations, economic geography, sociology, and planning, the regional scale has emerged as an organizing concept for interpretations of economic change. This book draws on the rich contemporary literature on the region but also addresses theoretical questions that preceded "the new regionalism." Geographers, such as Doreen Massey (1979) and David Harvey (1989); economists, such as Bennett Harrison (1994a) and policy-makers, such as Stuart Holland (1976) raised "the regional question" in the context of arguments about equity and social justice. They understood the regional question as a way of thinking about social relations in space and about the forces shaping people's opportunities and livelihoods in a world in which capital was increasingly mobile. Within this paradigm, space – particularly the regional scale – was problematic and so, the "regional question" spurred debate about how regional spaces were organized and to what purpose.

Normatively influenced perspectives on regions and the space economy have continued in the work of Massey, Harvey, and others. In addition, valuable work has been done on intra-regional fragmentation and its corrosive consequences for the citizens of metropolitan regions (Dreier *et al.* 2001). At the same time, however, regionally focused economic development policy, "the new regionalism," has addressed enduring questions about the region from a very limited vantage point. In particular, the region has been conceptualized in ways that limit our ability to ask and answer critical questions about how regional spaces are being re-made and for what and whose purposes. This book is intended to widen the scope of questions asked about the region as a central scale of action in the global economy. It is both a critique of the "new regionalism" and a call to return to the "regional question," including all of its "concerns about the nature, causes, and consequences of forms of regional distinction" (Webber 1982).

At the core of this book are case studies of two industries that rely on skilled, innovative, and flexible workers – the optics and imaging industry and the film and television industry. These industries maintain a knowledge base in specific regions – optics and imaging in Rochester, New York and film and television in Los Angeles, California. Both of these industries have strong networks of small and medium-sized firms. They also have major transnational firms (TNCs) that dominate global distribution markets. In each case, and not coincidental to our story, the TNCs operate within US corporate governance institutions. As publicly traded firms, their top priority is increasing shareholder value over the short term. The incentive structure within which these firms operate shapes their strategies vis-à-vis the small innovative firms in their industry network, the regions in which they reside, and the high skilled and less-skilled workers who produce and distribute their products. Our focus on US-based firms and regions circumscribes the story we have to tell but also demonstrates the continuing influence of nationally constituted rules on the capacities and strategies of firms operating in international markets. By extension, it raises questions about models of market governance and their implications for regions.

Our intensive research on photonics and entertainment media firms, both large and small, led us to question some basic assumptions behind the new regionalism and to develop an alternative framework for understanding regional economic development policy. This alternative framework is captured in a set of premises about firm strategies and regional labor markets that conflict with some of the taken-for-granted assumptions that inform much of contemporary regional policy.

Premise one: regions present firms with a set of strategic options along with production locations

One taken-for-granted idea that has been critical to the development of regionally focused economic development policy is that firms make a choice between "high road" and "low road" strategies as they respond to changing global production and consumption markets. In part, this idea stems from an interpretation of globalization as constructing conflicting choices. For the large transnational firms that are central players in re-making regions, however, a dichotomous choice between a high road and a low road doesn't adequately convey the available options. Our research indicates that TNCs frequently combine low road (cost driven) and high road (high productivity) strategies and also find ways to reduce the bargaining power of experienced and highly educated workers – aligning cost reduction with access to a high-skilled workforce. The capacity to strategically combine different location and labor force options attests to the political as well as economic power of these key players in the global economy and as our case studies demonstrate, within the region. It also raises questions about the way in which the actual processes of constructing markets and integrating production processes are portrayed.

In the popular literature, these processes of integration and linkage are captured by Thomas Friedman's contention that "the world is flat" (2005). In Friedman's new world, fostered by trade liberalization and deregulation, firms are price-takers and so getting the prices right is the single most important objective for firms wishing to become competitive players in the global economy. The role of regional policy is to "aid" firms in adjusting to a more competitive global economy and to reduce entry barriers for small firms so as to further increase competition, thus increasing productivity.

On the other side of the ledger, firms can escape price driven competition (at least theoretically) by producing value-added products and by continuous innovation. This "high road" strategy is central to regional economic development policy because the theory suggests that it can shelter the region from the consequences of cost competition among firms. Within this explanatory framework, the achievement of a more protected status in the global economy requires active regional policy. According to the new regionalist narrative, regional policy-makers must create the conditions within which firms can continuously innovate and add value to their products. Regions must provide research infrastructure, a positive business climate, amenities to attract talented managers, and, preeminently, a skilled and creative workforce. Regions that provide all the inputs to help "their" firms continuously innovate cannot be guaranteed protection from the unrelenting forces of global competition. The message is clear, however, that failure to make the public investments that will allow regionally located firms to pursue "the high road" leaves the region only the option of competing on the basis of cost.

Our research suggests that this understanding of the available options is not held by the managers of transnational corporations. Their perception of the options (both political and economic) is more accurately portrayed in the strategic management literature (Porter 1990) which presents a considerably different picture of the emerging global economy and how to succeed in it. From the strategic management perspective, firm managers must construct and secure competitive advantage, for example, by exploiting regionalized pools of skilled labor and finding ways to compete that will enable the firm to survive market volatility and surmount the drawbacks of competition based on price. From the strategic management perspective, getting the prices right is the wrong way to go. The corporate manager's goal is sustainable competitive advantage in an oligopolistic industry. Innovation plays a role but one subservient to the larger and more important goal of market dominance. This is a world in which merger and acquisition are critical tools and where the ability to shape the market through regulatory policy is central to constructing a firm that can reap the gains of opening global markets while significantly decreasing the risks of potential global competition.

In achieving the goal of sustainable competitive advantage, regions are important but, again, in relation to strategies and as a means to an end. In our industry cases, TNC managers lobby the region to provide, simultaneously, low costs of production, innovative capacity, and access to high-skilled labor. They look to regional policy as an instrumental vehicle to reduce their costs and risks (through firm-specific incentives), to provide the labor force they need, from entry-level workers to high-skilled professionals, and to provide the amenities and environment that will attract and keep TNC managers and high-skilled workers. Of course, few regions can succeed in this game and those that do, such as California's Silicon Valley, suffer the significant ills produced by diseconomies of scale and exacerbated inequalities.

Thus, our research on firm networks in high-skill knowledge-based industries indicates that the standard depiction of the newly defined role of regions in the global economy fails to capture the dynamics shaping firm behavior and its outcomes for regions and labor. It particularly fails in answering some key questions: Are there costs to creating conditions that firms can exploit in their drive to achieve sustainable competitive advantage in global markets? Who is defining what is meant by innovation?

Are there trade-offs involved in serving the needs of TNCs against those of small innovative firms? What does the TNC-centric model mean for the creation of sustainable regional economies? Is it possible to realize a learning region that builds a higher quality of life for everyone? These are among the questions not answered by the "new regionalist" conception of economic development.

To the contrary, current policies reflect a fundamental misunderstanding of the relationship between firms and regions in the global economy. The missing element is the question of power, exercised in networks, product and labor markets, and vis-à-vis the regulatory regime that sets the terms for inter-firm competition.

Premise two: power matters in firm networks

The new regionalist literature that currently dominates business and academic conversation rarely examines what have been central elements in the analysis of space making – economic and political power. Although there are exceptions, the analyses of regional agglomeration economies and firm networks have been missing any discussion of power relations, such as those between capital and labor, among firms with different political and economic capacities, or between firms and regional or national governance regimes, private and state-sponsored.

In our case studies, we particularly concentrate on labor, both as the subject of firm strategies and as important actors in collaboration with, and in opposition to, firm strategies. By bringing firm strategies and labor together, we reintroduce the concept of power into the analysis of the contemporary geography of production. We argue that political analysis is particularly important to the geography of the information or knowledge economy because so many firm strategies are aimed at altering the rules that govern production and labor markets (Holland 1976).

Unfortunately, in its lack of attention to power relations, and emphasis on trust relations and "soft infrastructure," the contemporary literature on regions and firm networks is afflicted by some of the same theoretical problems as the concept of social capital (DeFilippis 2001).

Networks of all kinds, including firm networks, are constructed around power relations. Networks encompass hierarchies of power or they wouldn't be networks. There would be no incentive for the more powerful members to remain in the network if they didn't disproportionately gain the benefits of network participation. Just as individuals "network" in order to promote their individual interests (rather than those of the network as a whole), so do firms. Networks can and frequently do take the form of hierarchies, with marginal benefit to the less powerful members.

A second important characteristic of networks is their exclusivity. There is no utility in belonging to a network if it does not keep people or firms outside its boundaries. In this instance too, the rewards of exclusivity disproportionately go to the more powerful members of the network who can control who is in and who is out.[1]

There are important examples of open networks, such as that supporting Linux, intended to contravene the exclusivity and control manifested in the vast majority of cases. Open networks are, however, the exception that proves the rule.

The neglect of the concept of power in regional networks is particularly problematic since one of the key agents implicated in the transformation of space in a global economy is the transnational firm. Because of their size, scale, and political-economic power, an understanding of TNC strategies is critical to any comprehensive understanding of spatial transformation, including the emergence and construction of production and market regions.

What is most notable about the TNC and something we examine in detail in our case studies is the nature of their attachment to the region. While we find that TNCs are dependent on regional pools of skilled labor and other production resources, we also see how their strategies and actions are defined by their access to global networks. Unlike their local suppliers of production inputs and innovations, it is easier for TNCs to escape the boundaries of the regional network and potentially to gain access to multiple regional networks. As a consequence of this ability to operate in but also across specialized industrial regions, they can exercise power over those firms that are captured in the regional net, and over regional labor markets, even those composed of highly skilled workers.

Also missing from contemporary theory about regions is an account of how more powerful firms exercise political and economic power at various spatial scales in order to shape the labor markets and production environments in which they operate. In devising strategies for achieving sustainable competitive advantage, TNCs are not limited to simple choices based on a set of locationally specific conditions which they must accept or reject. Instead, they actively shape the conditions in which they make choices through political as well as economic action at all geographic scales.

In an odd way the role of TNCs may be neglected because they are seen as dinosaurs, the relics of an earlier age of mass production. Certainly that was the message of Piore and Sabel (1984) in their seminal work, *The Second Industrial Divide*. Large corporations, with their stodgy bureaucracies and lack of ability to innovate, represented the past. The future was perceived to lie in small firms, clusters of trusting and cooperating entrepreneurs. Large firms and, especially, the TNC were identified with inflexibility, with the "organization man," the antithesis of the flexibly specialized, regionally-based entrepreneur.

In the 1980s, as the conversation about economic growth, innovation, and new production spaces shifted toward the regional scale, the key questions moved away from the power of the TNC and centered on whether regions were hospitable to entrepreneurial clusters of innovative, flexible small firms. The action shifted to within the region. Regional fortunes were measured in terms of endogenous factors – leadership, industrial adaptability, civic capacity. The role of the TNC was largely absent in this paradigm, perhaps because it raised unsettling questions about the limits of regional actors to influence the direction of regional economies.

There were some critical voices, however . . . Bennett Harrison's trenchant critique of the neglect of the role large corporations play in the global economy is even more true today than when he wrote about it in *Lean and Mean*. Harrison's insights about the continued power of large corporations in shaping and re-shaping labor markets and regional production centers to meet their needs were particularly prescient. He emphasized the compatibility of (industrial) concentration with the decentralization of production and, most importantly, pointed to the sources of decentralized production centers:

> Rather than dwindling away, concentrated economic power is changing its shape, as the big firms create all manner of networks, alliances, short- and long-term financial and technology deals – with one another, with government at all levels, and with legions of generally (although not invariably) smaller firms who act as their suppliers and subcontractors. True, production is increasingly being decentralized, as managers try to enhance their flexibility (that is, hedge their bets). . . . But decentralization of production does not imply the end of unequal economic *power* among firms – let alone among the different classes of workers who are employed in the different segments of these networks.
>
> (Harrison 1994a: 8–9)

Like Harrison, we do not find the TNC's power ebbing away as competitive entrepreneurs move into regional production complexes in a global economy. Quite the contrary. One of the key arguments in this book is that TNCs have adapted effectively to new challenges and opportunities so as to maintain and, in fact, increase their control over what is produced and how it is produced.

If we put the large TNC back into the contemporary regional picture, we can understand some of the apparent anomalies that commonly crop up in the literature on potentially innovative regional economies. The failure of a region to thrive is depicted commonly as a consequence of ineptitude within the region but may be a quite explicable outcome of different firm agendas and capacities to realize those agendas. For example, the absence of cooperation and the presence of knowledge asymmetries are depicted as inadequacies within the small firms network rather than symptoms of power differentials among large and small firms. From our perspective, these failures suggest the need for an explanatory framework that examines power differences as central to the dynamics of inter-firm interaction.

Our work tells us that TNCs and the small firms that supply them (and more importantly, provide the basis for innovation) do not operate in parallel universes – the TNC in global markets and small firms in the region. They come together, intersect, and compete for resources in the region. The power balance in that competition is highly one-sided and became more so, over the 1990s.

As we will demonstrate in the next chapters, the effects of this power imbalance are particularly visible in regional innovation systems and labor markets.

Premise three: labor skills are central to firm cost and innovation strategies

What becomes clear in analyzing studies of firm responses to trade liberalization, deregulation, and increased competition is that the labor force is the key element in firm location choices and in its strategies to achieve competitive advantage (Hudson 2001). In the contemporary knowledge economy, the search for skilled labor and creative capacities are central to firm strategies (Florida 2002a; Saxenian 1994). This is not news. Michael Storper laid out a labor theory of location in the 1980s that demonstrated the centrality of labor in location decisions as the relative cost of other inputs to production and distribution declined. What is surprising, however, is the lack of curiosity about the role of labor and labor skills in firm decisions, in a world in which labor and labor skills are highly differentiated and in which TNCS have considerable power to shape regional labor markets.

Our research and analysis foregrounds the role of labor and labor skills – the labor market is the key lens shaping our research. The search for labor skills is, however, understood in the context of firm strategies that are undertaken in the interest of positioning the firm in global markets. Choices about which labor markets to use and how to use them are not explicable in terms of simple, static economic calculations. They manifest strategies aimed at developing bargaining advantages with workers and regions over the distribution of risks and returns.

Two processes have changed since spatial analysts fixed on labor as the key factor in location decision-making. The first is the ability of TNCs to identify, locate, and use different labor pools, including skilled labor pools, to achieve different strategic purposes. The second is the ability to shape labor markets within regions to better meet the firm's strategic objectives and reduce its risks.

Skilled labor is not a unitary concept. Firms use skilled labor in different regional labor markets for different purposes. This has been noted by researchers who demonstrate the ways in which firms may be looking for brain-power rather than innovative capacity. For example, studies in emerging economies, such as those in Eastern Europe, India, and Turkey, with a supply of labor skilled in engineering and computer sciences, show that TNCs distinguish between the skilled labor they need for different strategic objectives, particularly their need for specialized skills and their need for innovation (Erbil 2006; Ionescu-Heroiu 2007).

The second development in strategic use of regionalized labor pools is an increased capacity to use *intra-regional* resources, both public and private, to obtain labor skills flexibly, in response to changes in market demand. This capability has always been present in the Los Angeles media entertainment industry but we found that it had also emerged in what are thought of as conventional labor markets, such as that supplying the photonics industry in Rochester. A combination of local and international outsourcing, adroit use of regional labor market intermediaries, and control of publicly financed innovation centers provides TNCS in Rochester photonics with a combination of flexibility with respect to high-skilled and semi-skilled labor and access to innovative capacity.

Grimshaw and Rubery (2005) provide insights into this process of intra-regional risk redistribution among firms. They describe how "unequal status among organizations" sheds a new light on how costs and risk are distributed among parties within a network and at the regional scale. In addition to transactions costs, power plays a role in how regional employment relations are structured – TNCs have more power to structure labor relations within regions, beyond the boundary of the firm.

Through market concentration and product line convergence, firms can create, albeit in a modified form, the lower risk conditions of the era of mass production and achieve economies of scale as well as scope. Through downsizing and the restructuring of local labor markets, including complex production networks, firms can transfer the risks of market volatility to the workforce and the small and medium-sized firms that employ them (Harrison 1994b). In these two ways – through strategic use of regionalized pools of skilled labor, and the re-construction of intra-regional labor markets – firms can reposition themselves in the global economy to secure the benefits of flexible production while at the same time reaping the rewards of more predictable mass product markets. As Henry Yeung argues, "Geographical scale has . . . become an important weapon in the continuous struggle between capital and labour in an era of accelerated global competition" (Yeung 2002).

Premise four: the role of the regional scale is becoming more important – as a source of subsidy and risk reduction to firms competing in global markets

In the 1970s, firms relied on the nation-state as a source of protection from the slings and arrows of competition in world markets. Partly because of firm initiatives to alter regulatory structures and institutions so as to create more opportunities for speculative profits, the protection afforded by national trade regulation is no longer available. Firms have not stopped looking, however, for sources of political protection from the risks attendant on competing in global markets.

Our research highlights how the regional scale has been singled out to absorb risks and costs for firms. Under the new regionalist paradigm, "regions" interact directly in the world economy. However, it is firms located in regions that have this capacity, not the places themselves.

The new regionalism does not adequately recognize the difference between regions and firm actors but, instead, obscures the boundary between the region and the firm. For transnational firms, the region is a convenient locus of action, relatively free of the onus of government accountability but, in the United States at least, still encompassing initiative, regulatory, and taxing power that can be put to the service of firm strategies.

That firms want to use regional capacities is manifestly apparent. The other side of the story – from the point of view of the regions – is equally important. What regions are experiencing is the disaggregated pieces of macroeconomic processes playing out unevenly across the nation-state – devolution of responsibility for social welfare and infrastructure and the consequences of deregulation and trade liberalization. The first of these has placed fiscal stress at the regional scale while the second has driven firms to search for scale economies. These processes have driven a wave of investment and disinvestment that lies at the heart of the regional inequality that has emerged since the 1980s. The pockets of deindustrialization and decline, the stars of the high-technology industries, the stagnant places, the growing places, the declining places, the old places, the new places, and the places remaking themselves for a new economy are only the symptoms of that process of investment and disinvestment.

The question then is to what extent regional economic development policymakers can choose a scale of action, independent of the exogenous realities of a macro-economy or the political realities of the city and the state. New regionalism is an effort to manufacture a scale – the region – in which local actors believe they can act effectively regardless of the political and economic realities operating on them.

The policy prescriptions proposed through the logic of new regionalism suggest that what is good for a regionally dominant firm is good for the region. Regional institutions, including universities and workforce training institutions, become the implicit and many times explicit partners of TNCs – providing subsidized research and development capacity, training for a skilled labor market, investment in industry-specific infrastructure – in an arrangement that neither recognizes nor accommodates for opportunity costs to regional residents.

The assumption underlying these investments is that public, private, and civic investments in the infrastructure that makes the TNC and its network of suppliers more competitive will simultaneously boost the economic competitiveness of the region. As our case studies suggest, this narrative fails to address the distribution of risk and costs playing out across places. Ultimately, a firm's success or the competitiveness of an industry does not necessarily translate into a sustainable regional economy.

Our research focuses on aspects of firm strategies vis-à-vis regions that, while virtually absent from academic analyses and the business press, were strikingly apparent in our conversations with business executives about how they see their strategic options. Their strategies to improve their competitive position specifically involved government intervention. Because there are no regional units of government, however, demands for assistance (with respect to favorable tax policy, for example) fall on cities and counties. They also fall indirectly on state governments because cities and counties look to the state for programs and tax policies that will enable them to respond to firm requirements. Also, because large transnational firms have more influence at the state level, they are able to lobby for policies, such as support for centers of innovation, or tax rebates to lower their energy costs in a deregulated environment, that are funded out of state monies.

Because of the decentralized character of many of these demands and their positioning within public–private partnerships, they are largely invisible to the citizenry. Ultimately, however, the mechanisms of government appear to be more important than ever to the competitiveness of firms in a regionalized global economy. Firms say that they need these mechanisms to reconfigure the competitive rules of the game – to change how markets function, to provide subsidies to support high-risk investments, to open new markets, to enforce intellectual property rights, and to create production spaces buffered from the give and take of democratic practice. They legitimize their demands within an argument that links firm innovation to regional competitiveness, and regional support for innovative firms to regional prosperity. Our research suggests, however, that these links are weak if, in fact, they exist at all.

To fill out how our critique of new regionalism emerged and led to a contrarian set of premises about contemporary regionalism we move through a set of empirically informed arguments and illustrate them with two critical case studies.

Our agenda: firm strategies, labor markets, and the regional question

While putting the concept of power at the center of our analysis, we approach the regional question through two lenses. Our first "lens" is that of firm strategies. The second is that of the regional labor market. These two lenses allow us to focus on how firm managers attempt to use regional resources to improve the firm's competitive position, how regional public and private sector leaders respond to firm demands, and the implications for regional economic sustainability. We begin by describing the strategic behavior of firms, whom we treat as active participants in shaping both physical and regulatory spaces.

Certainly firm strategies include locational choices and networking to achieve economies of scope and scale. But a grasp of the full range of firm strategies is required to interpret the contemporary geography of production. They also include political strategies to remake labor markets and to reduce market uncertainties and risks, strategies to promote policies that ease mergers and acquisition, allow a free hand in post-acquisition restructuring, or sloughing off onerous pension obligations, essentially allow firms to reassign risk to other economic actors – the workforce or the state. These strategies develop partners to shoulder increasing costs and mitigate increasing risks in global markets (Badaracco 1991). Ultimately, these strategies are intended to mitigate the firm's exposure to the increasing volatility of the global economy and create the basis for sustainable competitive advantage.

So, while firm decisions about how to organize production are central to regional outcomes, those decisions cannot be understood apart from the wider range of strategic options open to firms, including those aimed at changing how markets are governed.

Firm strategies, political as well as economic, are key to any thorough understanding of contemporary locational patterns and the relationships among places and within regions. In our research, firms employ strategies at *all scales of government* to construct markets and production spaces that will reduce risks and increase profits. For example, in the case of the film industry, the ability of conglomerates to operate at multiple scales and across multiple regions to change the risks associated with product markets has given them the ability to change the production process and their locational strategies.

From the broadest theoretical perspective, our examination of the regional question recognizes that the construction of a regional action space is simultaneously an economic, political, and imaginative project (Harvey 1990; Soja 1989).

Our approach has strong connections to the studies of industry restructuring that emerged during the 1970s and 1980s and to a re-awakened interest in the questions raised by firm decisions in response to political and economic as well as regional environments. That these questions remain a lively subject of interest and debate is suggested by Dicken and Malmberg:

> We need . . . a better understanding of how firms are being organized and reorganized; how internal and external power structures are configured and reconfigured; how business strategies are developed and implemented, as part of the dynamics of the wider industrial systems of which firms are a

part; and how each of these dimensions are "territorialized." This involves recognizing the nature of the firms not only as legally bounded entities and owners of proprietary assets (both tangible and intangible) but also as institutions with permeable and highly blurred boundaries.

(Dicken & Malmberg 2001: 346–7)

The industrial restructuring studies introduced methods to analyze how industries change over time, and theories about what moves them to change, what they produce, and how and where they produce it (Bluestone & Harrison 1982; Goodman 1979; Markusen 1985, 1987; Massey & Meegan 1982). They resonate with more recent analyses that emphasize the importance of examining process in attempting to interpret outcomes (Brenner 2004).

At the micro-level, there are a small group of researchers who have looked critically at firm strategies in relation to the risks and opportunities they face in the emerging global economy (Glasmeier 2000; Schoenberger 1999). Their work indicates that these strategies reflect a particular firm's culture and learning curve; they capture something that industry trend lines alone cannot reveal.

In our analysis of firm strategies, we treat TNCs as a special case because of their capacities and resources but also look carefully at the origins of TNC power vis-à-vis the region and regional labor forces. Our analysis of the transnational firm as an interested actor in national and regional environments contravenes conventional wisdom that TNCS are global actors who operate only in the global arena and "unlike real people, may exist in many places at once" (Greider 2003). This ability, to exist in many places at once, has led to the mistaken assumption that transnational firms represent the borderless world and exist beyond the reach of national politics. Our analysis of firm strategies, laid out in the next chapter, builds on a substantial literature that demonstrates how TNCs are shaped by national institutional environments, which both provide them with capacities and constrain their abilities to move freely through world markets. In this book we look at US-based transnational firms, not as unfettered free market actors but as products of a particular market governance regime. The US regime provides a valuable set of advantages and assets, particularly the ability for some firms to swallow competitors and achieve sustainable competitive advantage in an oligopolistic market, and the flexibility to move rapidly in response to changes in demand. The regime also constructs disadvantages, particularly creating unpredictable labor supply conditions. It is some of these disadvantages that US firms attempt to address through policy initiatives at the regional scale.

Firm strategies then are not simply a question of production location. In fact, location decisions may have become less important with the panoply of spatially consequential options open to large, transnational firms. What we attempt to do in analyzing firm strategies is to broaden the understanding of how firm strategies exercised at multiple spatial scales have consequences for regions.

The second lens we use in attempting to understand how the regional scale is being remade to serve the requirements of firms operating in a global economy is that of the labor market. Although the key role of labor in location decisions is generally acknowledged, there has been little analysis of what the centrality of labor in firm decision-making means for regions. The one exception is a literature that makes the case that firms are following skilled labor (Florida 2002b). In fact, this has always been true. If you want skilled financial analysts or skilled actors, you go to New York City. If you want skilled musicians, you go to Los Angeles or Nashville.

Our case studies focus on regional labor markets that depend on skilled labor but, in our cases, skilled labor still follows jobs in an industry. A cinematographer wanting more than occasional jobs may want to live in Eugene, Oregon but he or she still has to move to Los Angeles in order to pursue a full-time career. An entrepreneur in photonics may want to start his business in Boise, Idaho but will more than likely be drawn to Rochester, New York to obtain the machining and engineering skills that can enable his business to grow. Thus, the cases we examine emphasize the role of labor skills as critical to regional agglomeration economies.

In analyzing these two regionalized industries, we look at the intersection of labor demand and labor supply rather than separating them as is the conventional practice. We probe how the most powerful firms in the regional network exert power over labor intermediaries (including unions in the case of the media industries) so as to ensure that their needs for skills and flexibility are met first. We also explore how the discourse of a regionally based global economy feeds into inter-regional competition and how that competition is driven by coalitions of labor and capital. The result is the undermining of distinctive skilled labor markets in which labor has considerable bargaining power and the construction of regionalized industry labor forces which can be used more flexibly and cost effectively within and across regional economies.

What does this understanding of the regional question imply for approaches to regional policy?

Recently geographers have begun to show renewed concern over the question of the policy relevance and public policy applications of the research in economic geography and whether that research promotes better conditions for real people in real places (Lovering 1999; Markusen 1999, 2001a; Martin 2001; Massey 2000; Pollard *et al.* 2000; Storper 2001). Ron Martin makes the argument for policy engagement as follows:

> the improvement of socioeconomic welfare had to be *one of the primary aims* of the discipline: the essential motivation is to change the world not just to analyze it (see Markusen 1999). This means several things. It behooves us to expose and explain the inequalities and injustices that

contemporary economic-political systems routinely produce. It also requires us to interrogate and evaluate existing policies and policymaking practices to reveal their limitations, biases, and effects. And it means seeking to exert a direct influence on policy-making processes, at all scales, with the aim of producing more appropriate and more effective forms of policy intervention.

(Martin 2001: 190)

What, then, does our alternative perspective on the regional question imply for regional policy? In Section 3 of this book, we take up two key frameworks for regional policy and examine why they have been so limited in producing sustainable regional development. The two frameworks are regional innovation systems and learning regions. Both are founded in commonsensical truths: innovation produces jobs, and knowledge produces problem-solving, creative solutions, and new products. The problem we see is in the links made between innovation and knowledge creation and the ability to grow and sustain healthy regional economies.

The first "paradox" we examine is that around the concept of innovation. Process innovation and product innovation have different implications for job creation, the first leading to fewer, high-skilled jobs and the second, to job creation, at least for industries in the first phase of the profit cycle (Markusen 1985). Even when an innovation emerges in a region, the ability of that region to foster and take advantage of product innovation is determined by the answers to a key set of questions: 1) Who is interested in seeing the innovation come to market and why? 2) Who is interested in financing commercialization and where are they located? 3) Where are the skills available to produce the new product? So, the power of large firms to control which innovations come to market, the location of venture capital, and the location of product production all may be geographically distanced from the site of invention.

We also probe the "disconnect" between theories that advocate regional learning as a basis for regional innovation (and development) and the reality of inequality in regions organized around knowledge-based industries. The problems here are specialization of knowledge, project-based work, and high measures of labor segmentation, all of which create barriers for workers who want to develop career paths by learning and gaining knowledge in an industry over time. Ironically, regional innovation systems are more likely to be characterized by skill shortages than by a culture of continuous learning.

We examine the ways in which, while learning resources may be present, they may not be used in ways that create collective benefits. Rather they are selectively used to enable firms to compete in the global economy. We also look at how the concept of the learning region might be re-framed to focus on the value of the necessary pre-conditions and to emphasize inclusiveness.

Our final chapter attempts to "put the pieces back together," laying out some ways to think about regional economic development that take into account the central role of the labor force and its capacities for learning, and recognizes the role of power in determining how those capacities are distributed. We also recognize the limits of learning and labor force-oriented policies in constructing regions that provide for a high quality of life for all regional residents. Access to a learning environment and to education are necessary but not sufficient conditions to create healthy sustainable regions. While we advocate ways to realize the learning region in all its positive potential, we look at the possibility for combining the investment orientation, which is at the center of the learning region, with a commitment to regional policy that aims to alleviate inequality and its costs. Learning and labor force policies need to be combined with economic development policies that foster healthy regions through affordable housing policies, access to health services, and a collective commitment to a higher quality of life for regional residents.

2 Firm strategies

Resources, context, and territory

In our introductory chapter we suggested that the ability to understand how regional fortunes are affected by "globalization" requires a perspective that is both dynamic and multi-scalar. In this chapter, we lay out an approach to understanding firm strategies in integrating world markets that reflects our dynamic, multi-scalar perspective, with the goal of linking firm strategies to territorial outcomes, particularly at the regional scale.

Our starting point will be familiar to most analysts of firm strategies – it posits that strategies emerge from resources held by the firm (such as a specialized, skilled workforce, or unusually large capital assets) as well as from learning and experience over time. From that point, however, our approach diverges, incorporating a more contextualized and political interpretation of the resources that contribute to firm strategies, including managers' investment and location decisions.

A political interpretation necessarily involves asking questions about power. In thinking about firm strategies, power resides both in immediately available resources and in the power to act to alter conditions in the interest of the firm. So, while we accept the conventional view of resources as internal to the individual firm, we attach them to "capabilities," a broader concept denoting the potential to exercise power (Allen 2004). Thus, we part company with the conventional literature on firm strategies by looking at where resources and the capabilities that lie behind them are located. This entails looking beyond the firm to the private and public governance context in which firms develop the capability to exercise power and in which firm managers develop strategies over time. By contrast with the dichotomy drawn between state power and capabilities, and the realm of firm actions, we see these two spheres of power as mutually constructed, that is, firms construct state (as well as private) regulation governing markets and market regulations set the incentive structures that guide firm strategies. So, while we recognize that firm managers strategize and make decisions in response to market conditions, we also recognize that market conditions are not immune from firm strategies.

We have evidence all around us that firms are market makers, not just market takers. In the US, firm strategies to shape the markets in which they operate include "lobbying" policy-makers in regulatory agencies, action in the courts, the use of campaign contributions, and social networks to influence policy choices. One need only read about the US Chamber of Commerce and its executive director, Thomas Donahue, to recognize that the rules of the governance game are not taken for granted but contested terrain that large corporations attempt to shape to their advantage. The US Chamber of Commerce is the largest lobbying organization in the United States and spent $53 million in 2004 to influence the legal framework governing US corporations. Thus, our central assumption: firms actively construct market governance frameworks and, as a consequence, resources and assets – material and symbolic.

A political perspective on firms and markets is an old idea, going back to Adam Smith, but recently has re-emerged to change the way in which firm power and strategies are understood. This new interest in the political role of firms does not mean that firms are the only actors making and re-making the rules that govern markets but rather that they are critical because of their intense interest in and consciousness of how governance frameworks affect whether they can achieve their objectives.

While there has been more attention paid to the way in which firms and their key actors are embedded in institutions and networks, attention to the spatial and territorial implications of this understanding has remained within narrow bounds. For the most part, it has been used to buttress a conception of the global economy as made up of still distinct and bounded territorial systems governing capitalist economies, that is, nation-states. The "varieties of capitalism" approach is thus, primarily, an antidote to theories of market convergence.[1]

What has been missing is a more nuanced examination of how politically constructed, territorially based market governance regimes can be used creatively and strategically by firm managers to make and re-make investment, factor (including labor), product and distribution markets. To build our argument, we first look at the implications of understanding firm strategies exclusively as a product of the resources available to autonomous firms and their use of those resources over time.

Firm resources and path dependency as explanations for firm strategies

Two dominant theories have shaped the understanding of firm strategies, one based in the firm resources school (Penrose 1995; Wernerfelt 1984) and the second, emerging from Porter's (1998) work on strategic management. The original formulators of the firm resources paradigm, Penrose and Werner felt, recognized that particular resource positions constrain firm choices as well as providing firms with capacities relative to markets. For example, the maintenance of a highly skilled workforce in the firm – an internal labor market – gives it unusual capacities for process innovation but also entails costs, continual investment in the workforce and efforts to retain skilled workers. Despite their insights into resource trade-offs, the idea that resources represent a set of choices with consequences for firm strategies, received scant attention. This meant that the possibility for a more contextualized understanding of firm strategies was lost, at least for a time.

What *has* been recognized in the management literature is path dependency – the concept that firm resources and competencies evolve over time. The understanding of learning and adaptation over time is incorporated in theories concerning the path dependent character of firm growth (Nelson and Winter 1982). Path dependency describes "the cultural and administrative heritage of accepted practices built up over the course of the firm's history" (Berger 2006; Heenan & Perlmutter 1979). Berger describes this path dependency in terms of a "dynamic legacies model, the reservoir, or legacy, of resources that have been shaped by the past." By resources, she refers to the "stock of experiences, skills, human talents, organizational capabilities, and institutional memory" – not only material resources. A firm's legacy or heritage of practices, built on successes, failures, and learning underlies what Heenan and Perlmutter (1979) refer to as its "strategic disposition."

Of course, the strategic disposition of a firm is determined by the industry in which it operates. This is particularly important to note because of the dominance of manufacturing firms in analyses of industry strategies. Industries whose profits derive from distribution, such as retail or entertainment, have distinctly different strategic dispositions than product manufacturers for consumer markets or intermediate inputs. Industry characteristics are so important that they have been portrayed as trumping other sources of differentiation. What this misses, however, is how industry-specific requirements (as well as the learning associated with path dependency) are intermediated by territorially based incentive structures. So, for example, US-originating Wal-Mart has different strategies in international markets than German-based Metro or French Carrefour, though all are food retailers that operate internationally (Aoyama & Schwarz 2006; Brunn 2006).

The concept of strategies is valuable because it is commonly understood in the management literature as a product of resources internal to the firm and responses to external change in market conditions – technological change, the opening of new markets, policy changes, or increasing competition. In formulating strategies, firms bring particular resources to bear, such as management skill, and, in the more sophisticated analyses, path dependent learning (Berger 2006).

Firm strategies are, however, almost universally conceived of as firm specific. Even if strategy patterns, such as cost competition, are recognized across firms, they are represented as an aggregate of the decisions of individual firms rather than explicable in terms of a set of incentives creating a decision-making environment. The firm-strategies literature treats the firm as autonomous from the market governance arrangements which constructs the incentives within which it operates. So, for example, it would be as though Wal-Mart had no interest in the National Labor Relations Board regulations regarding the rights of employees to unionize, or in the legal contracts that define their obligations to suppliers. On a local scale, the meaningful regulatory framework may be land-use regulations that govern the location of "big box" retail. Under the autonomous firm conception, Wal-Mart must respond to exogenously developed regulatory standards. Its strategies are limited to attempts to deal with the problems created by regulation.

In reality, Wal-Mart is an active player in ensuring that regulatory frameworks favor the interests of the corporation.[2]

Although other interests may be represented and in some cases may prevail, firm managers are engaged in structuring the rules under which the firm operates and within which its strategies are formulated. Ironically, the rules that provide the firm with distinct advantages in the home market, where it has the most political "clout," may work to its disadvantage in territorial markets operating under other rules. This is the lesson that Wal-Mart learned in its unsuccessful entry into the German market (Christopherson *et al.* 2006).

So, while not rejecting the idea that individual firm managers develop strategies in response to changes in external conditions, we return to Penrose's initial inclination to understand resources as also posing constraints on firm action and thus potentially contextualized by incentive structures outside the firm. We examine the governance *context* or environment within which firms strategize and attempt to use in their strategic interest. Our perspective parallels that of scholars of technology and society (in their case the process of technological change and adoption) in questioning whether what is depicted as "exogenous" is actually a separate realm from firm strategies.

In our analysis we look at how firm strategies extend to the market governance framework that provides the incentive structure for individual firm actions. While markets are unpredictable and full of risk, firm strategies will include attempts to shape those markets and their governance structures so that they align with firm interests (Edelman 2004). In carrying out those strategies, they also reshape governance territories. Thus, strategic actions by firms are shaped by territorially bounded market governance regimes and also create new territories. Today, this is most evident at the global scale as firms lobby for frameworks governing trade, intellectual property, and the use of the oceans. It is, however, also evident at the regional scale where firms lobby for and create "zones" within which market rules differ from those of surrounding areas.

Firm strategies as emerging from market governance regimes

As trade liberalization and privatization of formerly state regulated industries has proceeded across advanced economies, the different approaches to and outcomes of market integration led to a reconsideration of the thesis that globalization naturally entailed convergence around a common set of rules governing factor, distribution, and labor markets. This discussion was both enhanced and muddied by a connection to changes in social welfare regimes, occurring at the same historical moment. Both processes were depicted as representing the decline of state influence over markets as a consequence of global competition. National differences in the process and direction of welfare state dismantling and industry privatization, however, raised questions about whether a single standard of global governance could capture what was occurring (Brenner 2004). In the end, what has emerged is a more politicized interpretation of market governance and social welfare regimes.

Gourevitch & Shinn (2005) for example, posit that "different regulatory policies concerning corporate governance turn on political differences among countries – on the interest groups that press for one set of rules or another and on the political institutions that aggregate preferences to produce policies" (p. 2). Of course, in real life we understand that policy differences are not only the result of "aggregate preferences" but of power exercised by key actors to shape public policy and the market governance system in their own interest. As we have already suggested, firms (through their managers and shareholders) are key players in creating and altering regulatory regimes and policies.

A conception of firm strategies that recognizes firms as embedded in a regulatory culture has been controversial because it contradicts conventional wisdom about globalization. The global economy is portrayed typically as an economy in which territorial boundaries have been superceded and in which "global" firms operate independently of the cultural or political strictures of territorial states (Friedman 2005). It is also a highly competitive economy in which firms have limited options. In the standard depiction of the global economy, local monopolies shrink as competition in product markets intensifies. Globalization forces firms to match pay to productivity and managers lose control over the wages they pay because more firms are vying for the consumer's dollar. These pressures are universal and lead to convergence in firm strategies across economies. Within this version of "the global economy," governments are prone to fail in making economic policy and only abstract unregulated markets get things right.

A wide range of empirical studies from different theoretical perspectives indicate, however, that considerable differences continue to exist among territorially differentiated economies – in production organization, sectoral strengths and weaknesses, equity investment patterns, and (significantly for this analysis) labor market practices (cf: Dicken 1998; Doremus *et al.* 1998; Jacoby 2005; Lazonick & O'Sullivan 1997; Patel & Pavitt 1997; Pauly & Reich 1997; Roe 2003; Whitley 1992, 1999). This continued diversity, combined with evidence of selective adoption and adaptation of practices originating in different economies (Lane 1998; Katz & Darbishire 1999), has lent support to a different understanding of how firms strategize and respond to the opening and integration of factor and product distribution markets.

Close studies of territorially based market governance regimes indicate that firms have particular resources in formulating strategies vis-à-vis global markets as a result of the governance regimes in which they emerge and develop over time. As Penrose suggests, however, these resources are not an unalloyed good. They exist in combination with a set of resource constraints that limit firm strategies in various ways.

A perspective that emphasizes how the context within which firms operate creates the resources that firms can use vis-à-vis markets is particularly interesting since it raises questions about the characterization of economies as "rigid" or "entrepreneurial." If the market governance framework not only inhibits firm actions but also provides firms with capabilities and resources, then the concepts of the autonomous firm and the antagonistic sphere of regulation lose their explanatory power.

By contrast, those industrial theorists who have attempted to understand the origins of – and process through which – market governance rules have evolved, explicitly recognize their embeddedness in social and political institutions (Schutz 2001 plus institutional sociologists). They emphasize that governance frameworks including public (government) policies "create constraints and incentives, rather than dictating firm behavior and that managers construct business strategies taking those constraints and incentives into account" (Dobbin and Dowd 1997: 502). And, "public policies influence corporate behavior by framing the competitive environment rather than promoting specific practices" (ibid.). We would add that firms actively engage in the processes through which these constraints and incentives are constructed.

Among the most important of the rules governing markets are those that regulate competition among firms. The necessity to control the degree and nature of competition is so central to the operation of markets that it must be "considered as much a defining feature of market economies as the existence of competition itself" (Berk and Swanstrom 1995). The very possibility of competition depends on the adherence to rules, as is the case with any game. That said, very different rules can be developed to regulate competition among firms and individuals and to define the limits and meaning of trust in a market transaction.

Dobbin and Dowd (1997) lay out a useful set of factors that affect the ability of firms to compete in factor and product markets. They include: 1) the ability to control price competition and market access through concentration (mergers, acquisition) or cartels; 2) the ability to accumulate capital and, so, to withstand market fluctuations; 3) access to state capital to reduce costs, pay for workforce training, and subsidize research and development; and 4) influence over direct policies (such as anti-trust) overtly regulating competition. These factors are present in all capitalist market economies but they are present in different forms and different combinations. Their influence on market processes is also differentiated by the presence or absence of powerful interests, such as unions, whose bargaining power is affected by competition policy and its ability to construct market power. If we want to understand politically constructed differences among types of capitalism, we need to look at how firms manage these four "arenas" of competition and what strategies they employ to construct sustainable competitive advantage. These system-directed strategies entail exercising power but they also entail compromise and coalition building. The nature of that coalition building differs depending on the strength of other powerful interests, particularly that of labor (Roe 2003).

For example, in so-called "coordinated" systems, typically exemplified by Germany, competition is regulated via strong intra-sectoral institutions and organizations, such as unions, employers' associations and credentialing bodies, in cooperation with the government. The risk-reducing and competition-enhancing advantages of concentrated ownership and private sector governance occur in conjunction with broad sectoral fiduciary responsibilities and exacting behavioral norms

The literature describing and analyzing varieties of capitalism has also provided a starting point for discussions of firm strategies in response to integrating world markets. At base, this discussion is rooted in the idea that national political and economic institutions, and the power and agency they construct, have a profound bearing on how private sector and state actors try to shape national and global institutions in a global economy (Hall & Soskice 2001). Or, according to Regini (2000), "preexisting institutions play a key role in shaping responses to exogenous factors by acting as a filter or intervening variable between external pressures and the responses to them."

For example, in short-term, shareholder driven systems, such as that of the United States, the regulation of competition and risk is carried out within adversarial institutions, particularly the courts and legislative bodies as well as in institutions such as regulatory commissions and private governance institutions. In this governance framework, significant advantages accrue to those economic actors who have the capital resources to pursue court "battles" over long periods of time and to influence legislation and regulation.

Our analysis of firm strategies focuses explicitly on the firm model that emerges within the Anglo-American "variety of capitalism" with its specific blend of capabilities, resources, and limitations. This firm "model" is shaped by particular competitive policies and relations between principals – shareholder investors and agents – the managers who are designated to act on behalf of the shareholders. The separation of ownership and control that characterizes Anglo-American firms is a defining feature of this model and has important implications for firm strategies.

The nature and extent of competition is, to a large extent, driven by the separation of control and ownership. Because managers act independently and potentially can risk investors' capital while protecting their own assets in the company, the Anglo-American governance regime provides for checks on managers' prerogatives. These checks affect the strategies managers undertake on behalf of the firm. The most important of these is the obligation, underpinned by tacit and explicit rules governing fiduciary responsibility, to meet short-term targets for sales, profits, and return on investment. As Clark and Wrigley (1997) lay out, managers respond to this restrictive governance regime by developing strategies that will insulate them from external scrutiny by financial analysts and investors and maximize the extent to which they can individually profit from performance-related salaries and bonuses. One key managerial strategy in this type of governance regime is to increase the firm's excess cash flow and capital assets, giving managers more prerogatives to respond to market risks and opportunities and protecting them from the risks of raising funds via bond or equity markets (Clark & Wrigley 1997).

What this example is intended to suggest is that the governance regime in which firms evolve and in which their owners and managers develop rules is critical to understanding how firm managers perceive and act on their options in product, factor, and distribution markets. In the Anglo-American model, managers in publicly traded firms in which ownership and control are separated make strategic decisions within a relatively clear set of parameters.

The literature on varieties of capitalism has been effective in making these parameters and their consequences for action more transparent. As we have suggested in the beginning of this chapter, what has been missing is a sense of how governance models arise from and are continually influenced by a bounded political territory and also have the ability to transcend territories to shape a global economy.

Territory as implicated in incentives and strategies

For the most part the literature on firm strategies has taken a narrow view of the territorial basis of the institutional context in which firm strategies emerge. Certainly the embeddedness of firms in a polity has been recognized (Jacoby 2005). And, there is a significant literature on the role of the state, harkening back to the earlier literature that reified the state as a territorial box, within which rules were formulated and enforced (Taylor & Asheim 2001).

What has been missing is a more nuanced analysis of how the power and capacities that originate in territories: 1) affect firm strategies in global markets; and 2) are shaped by firm strategies as they attempt to remake market conditions to their advantage.

The understanding of firm capacities as embedded in political territories but also existing in networks of powerful actors who use those capacities to act across territories has been neglected because the political-economic literature on the varieties of capitalism has emerged out of a nation-state based rather than scalar perspective on space and territory. In fact, political-economic "territory" at all scales is implicated in firm resources and strategies.

For example, territory is implicated in firm strategies to protect their markets from "exotic invaders." The recent departure of Wal-Mart from Germany after a nine-year effort to penetrate the German market testifies to the defensive capacities built into the German market governance regime. In this case, privately held retailers, who had long ago divided up the German market, were able to collectively "wait out" Wal-Mart until its aggressive investors lost patience with continual losses. While rooted in rules governing inter-firm competition, their strategy was bolstered by regulations governing everything from building codes and land use to labor practices. Essentially these territorially based regulations made it impossible for Wal-Mart to realize the capacities that enable it to dominate retail markets in the very different governance regime of the US Beyond defensive strategies there are other ways in which we can look at territory as implicated in firm strategies. One example is the way in which we understand the territorial (that is politically bounded) dimensions of spatial differentiation.

In concert with Dicken and Malmberg (2001), we distinguish the territorial from the spatial realm of firm activities, which is concerned with variations in the availability of resources and business opportunities as well as spatial distance. The territorial and the spatial obviously intersect but we are particularly interested in examining the political dimensions of the territorial because they also illuminate responses to distance and opportunity, that is, responses to spatial differentiation.

The concept of territory underscores the transitory and dynamic nature of bounded areas – highlighting the ability of actors to alter and define places and spaces and accounting for the relative and shifting roles of the local, regional, national, and global. Thus "territory" is redefined and restructured in response to the need to alter governance mechanisms as actors devise new strategies. For example, firm lobbying in favor of county government to displace the power of cities may reflect a strategy to develop production capacities in a less transparent and less regulated suburban context.

It follows that firm strategies to shape market conditions occur at all the geographic scales in which they operate. At the regional scale, firm actions reflect the incentives derived from national politically constituted market rules, such as those setting standards for labor contracts, while attempting to shape local conditions, such as the local prevailing wage for particular skills. In another example, firm managers may invest in one region and disinvest in another because market governance rules encourage mobility to boost short-term gains and allocate many of the risks of firm location decisions to the state and to individual workers. This capacity enables them to pit one region against another without incurring substantial costs or risks.

That said, however, firms have an interdependent relationship with regions, depending on the ability of regional governance institutions to provide them with the inputs they need to compete effectively. While the literature on regional innovation systems holds that this interdependence leads to trust-based relationships, our case studies as well as a body of other evidence indicates that interdependent firm networks are characterized by considerable conflict (Rutherford & Holmes 2006). The potential for conflict increases with the market power of large firm players. It follows that TNCs in oligopolistic national markets would have a different relationship with regional firm networks than large firms that are in highly competitive national and international markets. Market concentration means that the large firm has more power over suppliers, the workforce, and entrepreneurial innovators as well as, ultimately, over the regions in which it operates. Market power inequities lead to the potential for exploitation, exclusion, and conflict. As we will discuss in subsequent chapters, these conflicts work against sustainable regional innovation systems.

So, when we attempt to understand firm strategies vis-à-vis regions, we need to perceive the firm as embedded in, but also in dynamic interaction with, governance institutions at multiple scales. However, depending on the governance system from which they emerge, firms, including TNCs, will face constraints as well as have resources in responding to changing market conditions. Finally, to the extent that TNCs are in oligopolistic markets (also by virtue of their strategic action), they have capacities of scale, scope, and location choice that enable them to dominate and shape regional innovation and production systems. They have more power to shape the market and territorial conditions in which they operate.

Recognition of the role of power in firm networks raises questions about how the degree and nature of competition in an industry is determined. The economics literature answers these questions in an abstract fashion, for example, through theories of "contestable markets." A realist view of market governance, however, points to the myriad ways in which firms find formal (legal) and informal ways to reduce the costs and risks of competition. Large firms, especially TNCs, have the widest scope of action to sustain their competitive advantage.

To make this power more concrete, we look at how national governance models understand and regulate competition, and at what different competition "rules" imply for TNCs. We then look at what the context of firm governance means for firm participation in innovation systems and with respect to creative, skilled labor. In later chapters we extend these findings to look at how these constraints and resources are expressed at the scale of the region.

The transnational firm as a particular case

How do transnational corporations fit in this scheme of nationally-constituted governance regimes? Doremus *et al.* (1998) make a strong argument for recognizing the territorial basis for firm agency, even for firms whose locational choices and strategies transcend national territorial boundaries. In their analysis, leading TNCs internalize:

> both the basic national institutions and underlying ideological frameworks within which they remain most firmly embedded . . . the strategies and structures of corporations can and do change as they operate internationally, but only to the extent that those underlying institutions and ideologies permit such change.
>
> (Doremus *et al.* 1998)

At the scale of the nation-state, firms learn, adapt, and strategize in interaction with rules governing property rights, risk allocation, and, in general, capacity for action. In describing the process through which corporations govern, Danielson (2005) echoes Dicken (2003) describing how:

> Actions, reactions and inactions by all players in the system must be taken into account to get an accurate picture of the regime itself . . . if the decisions of corporate actors are indistinguishable from the decisions of state actors in terms of regulatory and social effects, then treating one as a "private activity" and the other as "regulatory" or "governance" activity will likely lead to more than ideological confusion. Such counterfactual characterizations may well result in significant misunderstandings about the way the transnational regulatory regime actually functions.
>
> Danielson (2005: 415)

When Danielson refers to "the system," he explicitly includes market governance institutions and points to the active role firms (in this case, transnational firms) play in constructing the rules under which they will operate. The active role of firms in shaping market governance is also and perhaps most effectively made in economic histories, such as those which examine the development of the corporate form in the US (Weber 1998; Berk 1994).

By definition TNCs operate across national borders, particularly to access resources (such as cheap labor) that they cannot obtain within their national economy. The assumption in much of the globalization literature is that TNCs operate apart from any nationally constituted market rules. They represent the unregulated "global" by contrast with the still rule-bound territorial state.

Studies of TNC strategies, however, indicate that they continue to be bound up with the capacities and constraints that distinguish the territorially based governance regime within which they emerged and developed. For example, if they are publicly traded, as most are, US transnationals operate within a set of distinct constraints that require increasing returns to investment over short time horizons.

On the other hand, however, they have considerable independent capacity to construct and reconstruct the conditions under which they produce goods and services. This firm capacity arises from the governance regime that describes their legal obligations and prerogatives, and the extent and nature of their accountability. Under US corporate governance "rules," firm accountability is narrowly defined, extending only to shareholders (and notably excluding employees or the public interest). And, large US-based transnational firms have unusual access to regulators and legislators at all levels of government because of the way elections are conducted and financed in the US (Gierzynski 2000). Thus large firms have considerable ability to influence the labor regulations under which they operate, the corporate governance rules to which they must adhere as well as their enforcement, and to shape public investments that may benefit their private purposes.[3] All of these capacities have implications for their exercise of power and coordinating capacity at the regional scale, both in the United States and in other investment sites.

Thus, the governance framework within which TNCs evolve shapes both their corporate resources, and ultimately, their strategies. As Doremus *et al.* (1998) describe, the incentive structure according to which their "success" is determined has developed within politically bounded territories. So, although they may operate internationally, the constraints, resources, and path dependent learning they bring to international strategies are rooted in territorial politics.

Second, with size and scale, and the designation as a global player or "national champion" comes the potential ability to expand into new and emerging product markets as well as the ability to leverage competing factor markets (labor markets, supplier networks, commodity chains). Both theory and empirical evidence indicate that transnational corporations have disproportional influence at all territorial scales and thus more spatial options and bargaining power in the global economy. Perhaps, ironically, they have more power to shape the national regime that governs their capacities in input and product markets and also to shape the regional conditions within which they operate.

So, when we try to understand the strategic agenda of transnational firms, we need to take into account: 1) the history of the firm and its legacy of investments and industry-specific knowledge; and 2) the constraints and capabilities residing in the national governance regime within which the firm makes key investment decisions and defines accountability; and 3) the relative power of the corporation to shape the market conditions within which it operates.

All of this is widely known. What we add to this picture is an analysis of what TNC power means for regions and particularly for regional labor forces. The ability of TNCs to instrumentally use regional resources is rooted in the dependence of regional labor and local fixed capital on the economic power of the TNC. This power has been enhanced as TNCs gain greater control of distribution networks that link local firms and labor forces to global markets. It has also been enhanced by TNC political and economic strategies to achieve market dominance via concentration.

To understand how and why TNC power is exerted in regional innovation systems, we need to parse out firm agendas in innovation-oriented industries and in firm networks and why large firms with access to global markets have different agendas than small entrepreneurial firms that are at the heart of creativity and innovation.

Firm strategies in networks and the mobilization of regional resources

One of the key strategies pursued by firms to achieve sustainable competitive advantage is the use of networks. Networks have an ambiguous relationship with territories, that is, with politically defined and bounded units. At the intra-regional scale, networks of firms that provide agglomeration economies are explained as a developmental form that can enable the region to escape from the exigencies of global capitalism – particularly the cost-driven inter-regional competition that results from trade liberalization and open "deregulated" markets. The success of these intra-regional networks, however, is often empirically linked to territorially based investments, such as in research centers, infrastructure, or labor force skills. Our case studies in the photonics and media entertainment industries demonstrate the significance of these investments, particularly with respect to labor force skills.

When firm networks are linked across transnational space, their dependence on political action is, however, almost completely obscured. Transnational networks are depicted as fundamentally anti-territorial because the network is disconnected from "the logic and meaning (of places)" (Castells, p. 412 cited in Sheppard 2002). The concept of networks is used, in fact, to reinforce the irrelevance of territory. From this anti-territorial perspective, network-based conceptions of the geography of contemporary capitalism actively confront the utility of scale as a descriptive mode, emphasizing the space extensive, horizontal and connective role of networks especially in the creation of extra-regional formations (Dicken *et al.* 2001).

Some network skeptics question the de-territorialized, de-politicized conceptions that dominate depictions of transnational networks and the way in which they limit our understanding of the geography of capitalism, how it is emerging and what forms it is assuming. In particular, they question the silence on power relations within the network.

These critiques point out that networks provide a way of eluding basic questions of agency and structure: "Networks are represented as self-organizing, collaborative, nonhierarchical, and flexible, with a distinctive topological spatiality" (Sheppard 2002). Critiques from a variety of perspectives and empirical research on actually existing networks have reintroduced the question of power and inequality into network theory, with important implications for how we conceive of the networks that are being constructed in global space (Graham 1998; Zook and Brunn 2006) and within regions (Christopherson & Clark 2007b; Grimshaw & Rubery 2005). As Sheppard (2002) cautions, "more attention needs to be paid . . . to the internal spatial structure of and power hierarchies within networks and to their considerable resilience and path dependence."

In actuality, the ability of firm managers to mobilize key actors, including unions and smaller firms, in networked "collective" efforts (for example to make claims on public resources) is a representation of power. In the case of the large dominant TNC, this capability goes beyond an individual firm's power to dominate or to coerce. It describes the ability to make the TNC's interests appear as the interests of the collective – the network, the industry, or the region.

When we look at firm networks, we see how they lend themselves to this manifestation of power, and how TNCs' interests are represented as collective interests, those of the network as a whole. Our case studies provide some examples of how this kind of power emerges and how it is mobilized. In the media entertainment industry, a belief on the part of small firms that they have no other option but to collaborate with the stronger players encourages firms to support policies that increase the mobility of media entertainment capital. Regionalism plays a contradictory role in these processes. The motion picture unions, for example, encourage state-level public investment in the industry through subsidies to production, understanding that those subsidies overwhelmingly work to the benefit of the transnational firms. At the same time, the regional union locals are party to national contracts and so want to see other states provide subsidies in order to maintain peace within the union. Loyalty to the region conflicts with loyalty to the (national) union. The transnational firm benefits from that conflict because unions will encourage state subsidies to the industry across regions. Thus the union fosters capital mobility even at its own expense.

In the photonics industry, regional economic welfare has been associated so long with the wishes and needs of the regionally based TNCs that, even when the managers of these firms say that they can no longer be concerned with regional welfare, public officials continue to place them at the center of regional economic development policy. Thus, the policy-driving idea that the TNC represents the collective interests of the firm network and the region may continue even when it imposes costs on entrepreneurial small firms in the regional network and on the regional labor force.

Beyond differences in power to define and control the "collective enterprise" lead firms in networks behave in different ways strategically vis-à-vis "their" networks, depending upon the constraints and capacities constructed by the corporate governance system in which they emerge. So, for example, US-based TNCs are more likely to develop vertical networks in which they maintain control not only of what is produced but of how it is produced and marketed. This need for control is a consequence of the lead (TNC) firm's need to act quickly in response to market change, to be "nimble." The absence of speed in response may endanger the ability of the firm to achieve rapid and continuous increments in shareholder value (Christopherson 1999; Eisenmann and Bower 2000).

Short-term return-oriented governance systems tend to redistribute risk downward to suppliers and the workforce via performance contracts and non-dependence and non-compete clauses. The transaction costs of subcontracting are arguably higher in these systems because of the need to start over with every contract. In coordinated systems, such as Germany or Japan, long-term relations among core firms and suppliers allow core firms the flexibility associated with vertically disintegrated production without the high transaction costs. They also require a different distribution of risk in which core firms protect their suppliers from going under during economic downturns and share the costs of worker training and of new technology adoption. These governance system characteristics are particularly valuable in sectors, such as the automobile industry, where quality control and continuous improvement contribute to competitive advantage.[4]

By contrast with the ideal of inter-firm relations in key manufacturing industries as realized in coordinated economies, the US governance system constructs another set of advantages. In sectors where a short-term "product" is the ultimate goal – the advertising campaign, the financial "deal," the television program, the technological innovation – a different model of inter-firm relations is used to reduce risk and contribute to competitive advantage. Highly specialized inputs are combined by specialists working together as an ensemble rather than a team. Logistical and coordination skills replace social relations (or more typically, work in combination with them) to ensure rapid market entry and "first-mover" advantages. In these situations, market governance regimes that encourage individual investment in specialized, technical skills and realization of those assets through individual returns in a variety of forms of compensation may be more effective in producing a quality "product." As Gereffi (1996) describes: in short-term "venture" projects, the firm is a venue for combining specialized inputs around short-term investment goals. It is not a learning organization but a staging organization. Interrelationships among firms are based on ability to provide specialized inputs (presumably at the lowest cost) not on the ability to provide continuous input regarding product quality. Trust is rooted in experience with an individual not with a firm and so networks are individual; they may operate quite independently of the firm's objectives.

Empirical evidence that networks of firms and individuals in regional innovation systems may be constituted differently under different governance regimes, for example, exhibiting different inter-network power relations among network members, reinforces questions about the explanatory capability of actor networks. As theoretical devices, actor networks are "designed to bypass the structure/agency distinction in social theory: actors derive their intentionality, identity and morality from the network rather than as independent agents" (Sheppard 2002). Actor networks, however, do not replace structural explanations, which inherently deal with power relations but rather *sum up* a set of interactions (Latour 1993). At the same time, evidence of different power relations tends to support social network analysis, which has historically been concerned with network positions and inequalities among network members (Sheppard 2002; Latour 1993).

Conclusion

We typically think about firms as groups of people who bring particular capacities into the marketplace; technical advantages, skills, and innovative ideas. Their strategies are focused on market conditions and success is measured post hoc. Firms that are profitable are assumed to have successful market strategies and become the subject of business journalists who peer inside their operations and the minds of management in order to discover their secrets.

What impressed us in doing empirical work on firms, their options, and actions in what is now a global market and production environment is how narrowly firm strategies have been conceived. The typical accounts look at time only in terms of trend lines and at space in terms of a series of location decisions. While institutional approaches, such as path dependency, are an improvement on the mechanistic view of firm action, they still focus largely on the firm as a free-floating entity which responds to exogenous changes in the market, either successfully or unsuccessfully.

This chapter has laid out a way of understanding firm strategies that recognizes how incentive structures in governance regimes affect networks, and how capacities created by governance regimes affect whether inter-firm networks assume more horizontal or vertical forms. We have also raised questions about power in firm networks, a topic we will take up in more detail later in the text.

Also obvious to us in our analysis of regions, networks, and governance was how little attention was paid to the most important collective resource in regional production networks: the labor force – its skills, and tacit and codified knowledge, and ability to flexibly respond to changes in market demand. Though the critical role of the labor force has been acknowledged by some analysts (Malmberg & Power, 2005; Hudson & Williams, 1999) there has been little work linking firm strategies to the labor force apart from the use of out-sourcing to reduce costs.

Recognition of what Grimshaw and Rubery (2005) describe as "unequal status among organizations" sheds a new light on how costs and risk are distributed among parties within a network. In addition to transactions costs, power plays a role in how regional employment relations are structured.

In the next chapter we contribute to the emerging discussion on how network relationships and labor markets intersect by examining how these dynamics take shape in a regional context. We believe a regional scale analysis can contribute to the understanding of inter-organizational relations around labor by illuminating the instances in which power and competition trump cooperation.

3 Labor markets and the regional project

Introduction

This chapter continues the discussion of the role of the regional scale in the context of globalization with a focus on the labor market. Labor, like capital, is an essential factor of production. However, we argue that labor and capital are reshaping two different scales of production – the supra-national global and the sub-national regional. While firms, and particularly transnational corporations, can choose among an international array of locations, what differentiates those locations are their respective regional labor markets. This differentiation occurs within and across national governance regimes that have long shaped the contingent contexts of transnational firm strategies at the national scale as described in detail in the previous chapter.

This chapter engages three elements of the conceptual model outlined in the introduction. First, we explain the process by which the regional scale becomes the dominant scale for innovation and production through the demand for (skilled) labor. This contrasts with the dominance of the global scale driven by the demand for capital. The story of agglomeration economies is critical to understanding and explaining the emergence of the regional scale and its role in the geography of production.

Second, we use labor market flexibility as an example of how and in what ways the state and transnational corporations, at the national scale, shape a landscape in which regional differentiation becomes profoundly uneven. In our case studies we illustrate why labor market flexibility benefits firms. Here we spatialize labor flexibility, adding the issue of scale to a developing literature on "the new pyschological contract" (Stone 2001).

Finally, we argue that inter-regional competition is furthered and reinforced by labor flexibility. The inter-regional competition benefits firms while exploiting regions. To demonstrate this we extend the discussion of agglomeration economies to a discussion of agglomeration diseconomies and the ways in which firms manage labor within regions.

Firms combat scale diseconomies through the relocation of production, the restructuring of work, or the redistribution of costs and risks. The region is implicated in all of these strategies and scale operates within and across processes of relocation, restructuring, and redistribution. The region has become a site to which transnational corporations transfer risks and costs to gain competitive advantage through the externalization of labor reproduction. Richard Florida uses the "Three T's" of technology, tolerance, and talent, to describe core characteristics of competitive regions (Florida 2002b). We argue that the core characteristic of competitive regions is a willingness and capacity to absorb and adapt to the "three R's" of shifting firm strategies – relocation, restructuring, and redistribution. The media and photonics case studies bring specificity to these processes within the region.

In *Global Shift*, Peter Dicken provides a list of five reasons for the growing importance of labor markets at the regional scale within a global economy where value-added is determined by creativity and knowledge. Those reasons include: 1) labor skills and knowledge; 2) wage rates; 3) labor productivity; 4) labor control; 5) labor mobility (or labor fixity) (Dicken 2003). These factors are regionally differentiated and distinct across labor markets. While most discussions of regional competition and innovation systems focus on labor skills and the relative fixity of skilled labor markets, the questions of wage rates, productivity, and control apply across industries and skill levels. In both case studies these characteristics shape the regional labor markets and the relative attractiveness of the regions to firms. The process of shaping distinct labor market regions is recursive and deeply influenced by the state, the dominant firms within the region, and institutions and intermediaries.

Labor is both a site of regulation – intimately related to issues of governance and political economy – and an actor in the governance process. In the region, labor both acts and is acted upon. Labor defines and is defined by the region it occupies. The idea of a recursive relationship between labor and place is not new and has been revived in the literature on labor geography (Herod 1997; McDowell 1997). What it means for labor to be active rather than passive in the context of dynamic firm restructuring, however, remains minimally investigated: "New Regionalist conceptions of the labour market pay inadequate attention to the fact that labour markets are socially constructed and embody relations of power" (Lovering 1999). Doreen Massey, in *Spatial Divisions of Labor*, described this idea by arguing that labor markets are relational – emphasizing how they are socially, economically, and spatially constructed (Massey 1984).

Managing regional labor markets: agglomeration economies and diseconomies

Geographers have long argued that the benefits of agglomeration economies for firms include mitigating the costs associated with externalizing transactions (Pred 1977). In other words, vertical disintegration can result in net lower costs in the context of agglomeration economies despite an increase in the number and complexity of transactions occurring outside of the firm (Scott 1988c; Storper 1999). Here the idea is that technological innovations in information and communication systems, and transportation shape a shrinking and flatter geography of production (Dicken 2003; Friedman 2006). This technological explanation, however, fails to explain why regions seem to matter in the global economy or why some regions seem to retain their "stickiness" and hold firms and industries in place over time – or why they fail to do so forever (Markusen 1996).

Looking at regional development from a dynamic perspective provides an analysis that focuses on transition and change, not simply growth or decline (Crevoisier 2004; Massey 1979). Much of the discussion of regional economic development has focused on the positive relationship between industry agglomeration and regional growth (Scott & Storper 2003). While the importance of the regional labor market, especially in the context of innovation and technology with specific skills, is not a new idea, it is unusual for labor markets to take center stage.

Most studies of agglomeration economies focus on innovation networks rather than regional labor markets, and on relocation rather than restructuring. These studies often document the capacity of places to function as incubators for a particular technology or production process. Researchers argue that agglomeration is, in part, the result of embedded cultural or historical trends in an area but the regional labor market is rarely identified as more than a component of these "cultural factors" (Cooke 2002; Saxenian 1989; Wolfe 1999).

Some of the best known examples of these agglomeration studies include the spatial clustering of high-tech industries in Silicon Valley and along Boston's Route 128. Saxenian (1994) attributes these success stories largely to the presence of research universities and the innovation capacity they foster. Other researchers also credit generalized historical circumstances as the key to agglomerations. Examples include the finance industries in New York City and London (Thrift 1994). Still other researchers identify natural resources as key determinants in industry agglomerations such as the steel industry in Pittsburgh and oil in Houston. National defense policy is considered a crucial factor in some industrial development such as the concentration of military bases in San Antonio or San Diego or the airplane industry in Seattle (Gray *et al.* 1996; Markusen 1991). While labor markets are essential in all these cases, they often remain peripheral or entirely unexamined.

The emphasis on firms rather than on labor markets, however, has not been universal. There are some studies that engage the symbiosis between the region and the labor market. Susan Christopherson and Michael Storper's study of the media industry in Los Angeles is one such example (Christopherson & Storper 1989). Katherine Stone's study of the steel industry, while not explicitly place-bound, is another example of an ongoing engagement of industry strategy and its effect on labor markets (Stone 1973). Walkowitz's historical study of Troy and Cohoes, NY is also a prime example of an industry study that places labor on a par in the analysis of the industry through his emphasis on the "company town" phenomenon (Walkowitz 1978). However, studies with a focus on the regional labor market more often analyze employment change or worker displacement rather than the ongoing interactions between firm strategies and the regional labor market (Beneria & Santiago 2001; Pollard & Storper 1996).

In most regional case studies, the agglomeration economy is central to explanations of competitive advantage. Similarly, agglomeration economies provide the theoretical foundation of regional economic development policies focused on regional innovation systems, industry clusters, industrial districts, knowledge economies, and learning regions. Much of the popular work on economic development in the US in the last fifteen years rests on the idea that proximities produce distinct economic advantages (Florida 2002a; Krugman 1991; Porter & Stern 2001).

The MIT Dictionary of Modern Economics defines agglomeration economies as "cost savings in an economic activity which result from enterprises or activities locating near one another" (Pearce 1992). At their core, agglomeration economies are primarily a set of advantages derived from the physical proximity of a set of firms who share factors of production that would be more expensive (in terms of production and/or reproduction) if individual firms had to procure or provision each asset fully and separately. Through proximity, firms share the costs of factors of production.

Whether discussing regional economic development issues in terms of agglomeration economies or regional innovation systems, labor markets play a key role. Spatial economies derived from labor market pooling become more important as technological innovations reduce the costs of transportation and communication over larger distances (Fujita *et al.* 1999). As the advantages of geography are mitigated by technology and global options, the very fact of labor's relative immobility is increasingly important in understanding the economic benefits of individual regions to firms. Thus, a discussion of why regions matter shifts from a discussion of transaction costs to a discussion of skills and innovation.

Increasingly, evidence indicates that the labor market, as a key bridge between regions and firms, is critical to articulating why innovation and restructuring do not consistently produce regional growth (in wages and job creation) in a globalized economy (Cowie and Heathcott 2003; Rutherford and Holmes 2006). In many regions there is evidence of stagnant and declining wages, erosion of employer-provided health and pension benefits, and of cities and local governments struggling to provision basic services and infrastructure (Peck 1996; Osterman 1999; Stone 2004). These "disconnects" between production and place underscore the need for a broader understanding of firms, labor markets, and regional development. However, the connection between industrial agglomeration and the regional growth is predicated on an almost synonymous understanding of "region" and "industry" (Markusen 1994; Gray, Golob et al. 1996). This fluidity between the conceptions of place and production has grown in recent years and now extending beyond discussions of agglomeration economies to recent articulations of regional innovation systems (Morgan 2004).

However, agglomeration economies are not a single set of processes, nor do they work only in one direction. The same processes that produce agglomeration economies start a cycle of diseconomies in motion. Because places provide the benefits of shared costs, they become more attractive to firms and new firms enter the market and existing firms expand. These expansions and new entrants create new demand for localized, industry-specific inputs. The increasing demand subsequently bids up the cost of the very assets whose availability and affordability made the region attractive to firms in the first place. The diseconomies that lead to regional decline are often blamed on regional actors rather than understood as an almost inevitable flipside of the growth process celebrated in the agglomeration economies model. "By this means, regional problems are conceptualized, not as problems experienced by regions, but as problems, for which, somehow, those regions are to blame" (Massey 1979).

Although theoretical explanations and public policy discussions largely sidestep these cyclical effects, diseconomies are familiar to regions, as well as firms. The production landscape is strewn with examples of places that have experienced both the boom and the bust cycles in local markets. The upswing in prices, of land and labor, in Silicon Valley in the 1990s is but one example of diseconomies of agglomeration. Ottaviano and Puga characterize the diseconomies of agglomeration in their assessment of the "new economic geography" with particular attention to the question of increasing wages and firm strategies to avoid them. They argue that, "if equilibrium wage differences are not eliminated by migration, they act as a dispersion force by increasing production costs for firms producing in locations with relatively many other firms" (Ottaviano & Puga 1998).

The Silicon Valley experience serves as an example of diseconomies of agglomeration in labor markets. It also illustrates the ability of firms to strategically adapt to diseconomies. For example, in the Silicon Valley case, firms created alternative compensation systems to mitigate the immediate upward pressure on wages. Software firms also developed on-site perks for workers such as recreational opportunities and restaurants. These strategies paralleled the development of pension and other non-wage employee compensation strategies in the 1940s to attract workers while wage regulations were in place. In the optics and imaging case study, early-twentieth century corporate paternalism included perks such as on-site free optical care for employees.

The cases of photonics and the media industries illustrate the capacities of transnational firms to shift gears in response to diseconomies and thereby maintain competitiveness. While this adaptability is remarkable, it is also clear from empirical examples that the redistribution and restructuring rarely, if ever, work to improve the compensation of workers or the condition of the region. Long-term regional specializations, like those in Los Angeles and Rochester, provide a clear site for viewing the progress of restructuring in the region.

In our case studies of optics and imaging and media, firm strategies are obvious in the tension between these cycles of economies and diseconomies of agglomeration. As profit-maximizers, firms inevitably look for ways to avoid diseconomies or negative spillover effects. Transnational corporations are more adept at these strategies than small and medium-sized firms because they can use inter- and intra-regional competition to their advantage.

Firm strategies related to relocation, restructuring, and redistribution appear in and across regions as responses to mounting and/or cyclical diseconomies. Firms restructure as a response to exogenous changes in markets and prices. Successful firms adapt to global competition, technological change, or changes in trade policy (Schoenberger 1999). How firms shift strategies and capacities in their product markets – new product introductions, new advertising schemes, new management – is the focus of many a business school case study. How firms influence the markets for their inputs – labor, real estate, research capacity, and capital – is often neglected. The region plays a significant role in these strategies and the construction of political and economic power that makes negotiating across scale possible.

Labor market flexibility

By contrast with the static orientation of most case studies of regional innovation networks, the case studies in this book identify dynamic fissures in the production process which exacerbate existing inequalities and reproduce uneven development. We identify three distinct processes that produce gaps in the distribution of regional growth and all three exacerbate inter-regional competition and patterns of uneven development.

The first process is the largely unrecognized political-economic power of transnational corporations in regional firm networks and how that power shapes the allocation and distribution of factor inputs and resources within regions. This process was discussed in the previous chapter. The second process is that of increasing labor flexibility and the subsequent erosion of wage rates and skill specializations within the region that is discussed in this chapter. The third process is a shifting emphasis in public capital investment away from subsidies that promote infrastructure and education and toward the subsidization of innovation capacity for firms and industries. This process is discussed in detail in Chapters 6 and 7.

The dynamic processes of adaptation and agglomeration are often approached, in empirical case studies and in theoretical discussions, from a static position, as snapshots rather than as systems. This tendency is particularly evident in the research and thinking about regional labor markets and the recursive and dynamic relationship between firms and the local labor market, the institutional infrastructure that supports it, and the governance regimes that shape it. In these case studies, we approach the photonics and the media industry by focusing on the new forms of flexibility evident through the progressive reconfiguration of the regional labor markets.

These case studies take a broader look at the process of increasing flexibility to ask not whether regional labor markets have become more flexibly specialized – an argument consistent with current understandings of innovative industries – but rather how the industry has become more flexible and how that flexibility – territorial, regulatory, strategic – has reshaped regional labor markets and the process of inter-regional competition.

The conversations about flexibility and competitiveness are complex, with significant implications in terms of causality and intent. They are also inherently political. While much has been written about the potential for a new, flexibly specialized service-based economy and much anxiety has emerged around the drama and pain of restructuring (Cowie & Heathcott 2003), these are not mutually exclusive stories. The terrain of production organization has changed in significant ways and new forms of "flexibility," particularly production flexibility, are central. A variety of strategic choices were identified as flexibility by Piore, as well as others, in the early 1990s.

> The reforms are generally characterized by those responsible for them as an effort to "increase flexibility," but this term is subject to a variety of different interpretations by those who employ it . . . Some executives are seeking to reduce costs by forcing their subordinates to bear them, through lower wages in the case of the workforce and reduced profit margins in the case of subcontractors, vendor, and other external business collaborators. Other executives seek to improve efficiency by reforms in the nature of the collaborative relationship with subordinates: generally, they are seeking more flexibility to adjust quickly to the shifting business environment . . . Typically, the flexibility they are seeking is built into the productive apparatus by pre-programming automated equipment and cross-training workers to operate on several distinct products. Thus, it is a kind of flexible mass production as distinct from flexible specialization, where the production set is open-ended. Nonetheless, it is very different from classic mass production and from simple cost-cutting tactics within the traditional production strategy.
> (Michael Poire in Sengenberger *et al.* 1990)

In the contemporary period flexibility strategies have added new dimensions. Flexibility has gone far beyond work hours to encompass legal and regulatory prerogatives. It also entails the development of new institutions and relationships to the public sector. In the case of the labor market, it often appears that it is the "cost cutting" rather than the "efficiency" model that motivates firm strategies. Small firms themselves, particularly in the Rochester case, indicated that the squeezing of subcontractors through an annual "double digit productivity" strategy was par for the course when working with larger corporate clients. This strategy is yet another example of what Bennett Harrison called "the dark side of flexible production" (Harrison 1994a). The impulse behind these newer flexibility strategies is often profit maximization but not necessarily higher productivity (Christopherson 2004). This shifting corporate imperative raises critical questions for both public and private institutions asked to adapt to the new corporate model and accommodate to changing competitive firm strategies.

"The new psychological contract"

The "old employment relationship" characterized by internal labor markets and long-term employment is familiar to anyone acquainted with American popular culture and Homer Simpson's job at the nuclear power plant or Mr. Brady's job at the architecture firm. The new model is better characterized by the liminal employment status of characters on *Seinfeld* and *Friends*. The transition from the old model of employment to the flexible employment model, described by terms like "project based" and "contingent," has been subtle and partial. Subtle too are the impacts of this transition on regions and the spatial organization of production. Rhetoric about work has recently shifted focus from mass production and distribution to "knowledge work" and innovation. John Kenneth Galbraith recognized the implications of this move toward specialization nearly forty years ago:

> The real accomplishment of modern science and technology consists in taking ordinary men, informing them narrowly and deeply and then, through appropriate organization, arranging to have their knowledge combined with that of other specialized but equally ordinary men. This dispenses with the need for genius. The resulting performance, though less inspiring, is far more predictable.
>
> (Galbraith 1967)

The process of redistributing risk through reorganizing work is complex. Unfortunately, adding the dimensions of technology and geography to the equation does not simplify the analysis. Though daunting, the task of understanding the implications of flexibility for the economic sustainability of industries, and the regions in which they locate, is important for forming functional regional policy. Understanding why firms look to flexibility as a production strategy provides insights into whether regions must concede to such forms of work organization in order to remain competitive in a regional world. Scott and Storper outlined three major types of labor market flexibility that firms pursue in the context of flexible accumulation in their 1990 article: "Work organisation and local labour markets in an era of flexible production" (Storper & Scott 1990). First, firms attempt to individualize the employment relationship: for example, moving away from collective bargaining arrangements. Secondly, firms work to improve internal flexibility through strategies like job sharing, project teams, and multi-skilling. Thirdly, firms move towards external flexibility characterized by the use of peripheral work arrangements and workers (contingent and contract labor) (Storper & Scott 1990). These flexibility strategies are at the heart of the analysis of many researchers in labor and industrial relations (Benner 2001; Clark 1989; Van Jaarsveld 2002).

Labor flexibility involves both the redistribution of production and the reorganization of other work activities. Flexibility strategies result in both inter- and intra-regional shifts in production location. In one strategy, firms subcontract or outsource both production functions and professional services outside the formal firm boundaries to "independent" suppliers. Restructuring also involves contingent or contract workers, provided by temporary employment, working on site. In this case, the reorganization of work is legal rather than spatial. Downsizing and layoff strategies mean more than "job loss;" the reality is closer to "job redistribution." The restructuring strategy shifts work beyond the formal boundaries of the firm while the work (and workers) remain within the control of the firm through "networks." Proximity plays a key role in maintaining control in production networks – including temporary employment firms, captive subcontractors, to regional research centers.

Although frequently thought of as losing jobs to other countries, labor flexibility often results in a redistribution of work within or between regions (Harrison 1994b). Jobs are relocated in terms of their relationship to the boundaries of the firm and the strength of regional firm networks – hence regional specializations become crucial to ensuring redistribution of "peripheral" jobs within the region (Schoenberger 1999). Firms strategically retain employees in their "core competencies" – workers who are scarce and have technical expertise or workers involved with intellectual property or proprietary knowledge – while outsourcing other functions. While job tenure and stability seem to be decreasing for core and peripheral workers, core functions require the legal protections of a direct employment relationship while peripheral functions only present the firm with liabilities (workers' compensation rules, employment discrimination protections, etc.) (Stone 2004).

Firms mitigate variable costs by externalizing employment and reducing the number of permanent, direct employees. The progressive outsourcing and subcontracting of production and services previously internalized (e.g. routine clerical or janitorial work and customized producer services) is also an element of the vertical disintegration of transnational firms. By shifting workers outside of the firm's formal boundaries – through subcontracting, outsourcing, independent contracting, or the use of temporary employees – companies shift liability for those workers outside of the firm and increase the predictability of their labor costs (Harrison 1994a). By shifting the risks of employment to private for-profit labor market intermediaries, individual workers, and the region, firms reduce exposure to variability as well as legal liability. That variability can be pension legacy costs, rising health care premiums for employees, workers' compensation claims, or equal opportunity lawsuits. A generation of human resources professionals has developed management strategies to avoid these costs. Again, in response to shifts in governance and regulation, firms do not stand still; they adapt.

Transnational firms are able to maintain control of production without directly employing the workers or owning the subcontractors. Increasingly subcontractors report that large firms squeeze their profit margins, demanding annual double-digit productivity increases. For small suppliers this compromises their capacity to make investments in their own competitive futures. These strategies belie a general emphasis on short-term advantages as more firms and managers evaluate their success by quarterly criteria (e.g. share prices, short-run production costs, price/earnings ratios) rather than long-term performance measures (e.g. product innovation, long-term fiscal viability). A recent IBM commercial featured a psychiatrist advising an anxious corporate executive patient to consider increasing productivity rather than relying on incessant, neurotic cost-cutting strategies. The thought clearly takes the fictional executive by surprise.

These production strategies, organized around vertical disintegration and flexible employment, have specific spatial implications. "Just-in-time" flexible labor market strategies rely on the availability of skilled workers within the regional labor force – a localized pool of qualified labor. Allen Scott provides a basic description of the economic relationship between vertical disintegration and increased labor flexibility,

> First, vertical disintegration (due primarily to instabilities in production and exchange) is reinforced where employers seek to externalize their consumption of selected labour inputs and thus to head off possible internal upward drift of wages and benefits. This strategy is especially favored among employers with a core of skilled, high-wage workers who also have a demand for various low-skilled types of work.
>
> (Scott 1988a)

Scott goes on to describe the increasingly dual character of labor markets and argues that firms and workers are attracted to agglomeration economies in order to mitigate their respective costs of turnover and job/worker searches. Paul Krugman points out that this parallels the historical phenomenon of the "company town" but functioning with a local employer oligopoly rather than monopoly. Krugman argues that firms – particularly transnational firms – prefer to achieve labor pooling through "agglomeration centers" rather than by dominating a local economy.

> Firms would like to convince workers that they will not try to exploit their monopsony power, so that they can attract workers to their production location. But the only credible way to do this is to have enough firms in the location that there is an assurance of competition for workers. The commonsense idea that firms would like to have a company town in which workers could be exploited is right; but the point is that workers will shun

such towns if they can, so that firms will end up finding it more profitable to locate in agglomerated centers that are not company towns.

(Krugman 1991)

The regional flexible employment strategy works because the specialized labor market is constantly churning, not employed or unemployed in response to a single firm's fortunes (Cappelli & Neumark 2001). This churning insulates workers from the impression, if not the reality, of being captive employees.

For the flexible employment strategy to be cost effective, there must be a reliable pool of competent and technically capable workers in the region. Similarly, the "just-in-time" production and distribution system depends on reliable subcontractors (Carnoy *et al.* 1997; Florida & Kenney 1990). As a consequence of the need for skilled and available workers, firms face two, often competing, dilemmas when crafting a regional labor market strategy. The first is to maintain control over variable labor costs (e.g. wages, benefits, pensions, legal claims) and to reduce labor's claim on total profits (e.g. wages, stock options). The second is to obtain access to pools of skilled labor without incurring the costs of purchasing or producing labor skills in a flexible employment world. This employment arrangement presents challenges for both firms and workers:

> These new work systems demand substantially more from employees than did traditional arrangements. Employees need more skills, particularly team-related behavioral skills, to succeed in these new systems. And many of these skills can be provided only on the job, by the employer . . . The contradictions associated with these new systems for organizing work turn on the fact that the needs they generate seem to go in the opposite direction from the trends being introduced in the employment relationship. Thus, while new work systems seem to require greater job security, the reality seems to be that job security has declined. In addition, the new ways of organizing work require more employer training, but the incentives for employers to provide that training are reduced . . . Because reductions in the length of time that employees stay with a company reduce the period of time over which the employer can capture the benefits of that improved performance, they greatly limit the company's ability to provide that training in the first place.
>
> (Cappelli 1997)

Transnational firms have a growing strategic interest in finding, developing, and maintaining regional pools of skilled workers. This strategy both counters the costs of the recruitment and retention of specialized employees and allows firms to maintain multiple regional production centers, thus insulating them from the increased demands of workers. The flexible employment strategies produce a growing group of skilled employees who do not receive a commitment from firms and are hired explicitly on a contingent, contract, or project basis. The high technology industries in the US shifted to this model in the mid-1990s with a number of intermediaries, like WashTech and Webgrrls, emerging to manage job matching. The rise of project-based work in high-technology fields is an outgrowth of this effort by firms to maintain both flexibility and access to highly skilled labor (Christopherson 2002).

What is sometimes lost in the discussion of flexible production and work organization is the need both for specialized skills as well as a need for flexibility. It requires a new paradigm to meet the expectations of firm or industry-specific production or development skills without a firm commitment to long-term employment. In a flexible employment world workers have every incentive to develop generalizable skills while firms want workers who can meet specific, and discrete, needs. Workers want to be broadly employable while firms want only to hire for core tasks. The new paradigm must be capable of dealing with that inherent contradiction in the labor market between the demands of the employers and the incentives for workers.

These opposing interests create a role for the region as a site of a specialized labor market, not just a site of industry specialization. As firms need skilled workers but do not want a long-term employment relationship with them a disconnection emerges. This contradiction – the need for firm-specific skills and the desire for a flexible employment relationship – creates an increased dependence on the skills and capacity of the regional labor market as a whole. In this arrangement, proximity and place matters. The perceived character of the regional labor market becomes a significant locational advantage or disadvantage in inter-regional competition for transnational production. Regions with skilled labor markets may have an advantage not simply in competition with other regions, but in competition with firms for the benefits of growth. One important question for regional policy is whether an increased dependence of firms on skilled regional labor markets means places have more bargaining power in the new economy.

The tension between embedded labor markets and the locational flexibility of transnational corporations plays out through myriad tensions and contests within and across regions.

> On the one hand, the increasing capability to span boundaries and borders that networking affords to business would seem to have tilted the playing field decisively against locally elected and appointed economic development planners, vis-à-vis the plant location managers of the multilocational companies at the hubs or apexes of the networks. Yet, at the same time, precisely because the networking principle allows concentrated business organizations to coordinate operations across an ever more dispersed field of play, more decentralized production becomes increasingly feasible. But then it follows that, paradoxically, the comparative attractions of different locales actually take on an enhanced significance for industrial location.
>
> (Harrison 1994a)

The redistribution of risks and costs of employment and labor reproduction and the reorganization of work produces a labor market strategy with a conflict at its core – between increased flexibility and demand for firm-specific skills. This combination of firm strategies also conflicts with the process of relocating production. Flexible employment strategies depend on location. The presence, or absence, of an adequate regional pool of skilled labor is a fundamental ingredient in the flexible employment strategy.

Inter-regional competition and the erosion of advantage

While the bargaining power brought by the capacity to relocate and restructure production processes belongs primarily to firms, it is not an unmitigated power. The ability of firms to negotiate with labor using regional competitiveness (and inter-regional competition) as the core of the debate is predicated on the presence of a system of governance that makes such strategies viable and economically advantageous (Jonas 1996). Spatial relocation is not the only labor market strategy available to firms nor is it the only one that firms employ. Firms may choose to use restructuring and redistribution strategies, *in situ*, to ensure desirable labor market conditions (Barnes *et al.* 1990). The challenge they face is not simply one of the adequate flexible labor supply but one of sufficient labor control.

> The search for a new labor control "fix" (following, say, technical change in the labor process or a shift in the local labor supply) can trigger either relocation – a spatial strategy for engaging with a new labor supply – or changes in an industry's local labor market relations – an *in situ* strategy for engaging with a new labor supply. In a sense, labor control considerations bear on the costs of restructuring in place versus restructuring through space.
>
> (Peck 1992)

Labor control influences transnational firm strategies in the process of labor market restructuring (Peck 1992). And labor control – the extent to which governance regimes favor the firm in the employment relationship – has historically been the primary purview of state and local regulation in the US and thus highly differentiated by region (Befort 2003; Stone 2004). While firms influence the governance and organization of both their product markets and the markets for their factors of production, the regional labor market is the best illustration of firm strategies at the regional scale.

Our case studies of the optics and imaging industry and the media industry provide strong examples of how transnational firms strategically use their position of power both within and across regional flexible production agglomerations to reduce the cost of creating and reproducing a high-skilled workforce by transferring those costs (and attendant risks) to the public sector and individual workers. Our case studies also demonstrate how potentially mobile firms use the specter of inter-regional competition to control and "socialize" high-skilled workers to their strategic agendas – even when those agendas undermine the power of the workforce and negatively impact the regional economy.

Firm relocation strategies are designed to extract concessions from regions. They are too blunt a tool to be associated with an explicit or strategic reorganization of work processes. Generally relocation strategies involve an effort to bid down the wages of workers, counter the influence of organized labor, and demand subsidies and benefits from municipalities and states in the form of a now vast and ever-expanding menu of tax rebates and economic development packages. For decades firms have disciplined both workers and places and maintained control over labor and space using relocation strategies.

> The high-tech industrialists have, for example, pressed hard for reductions in state taxes and burdens imposed by other programs and regulations originally sponsored by the labor movement. And they have heavy-handedly suggested that if their demands are not met, they will move their firms to states with a more favorable business climate – a threat that borders on industrial blackmail.
>
> (Piore & Sabel 1984)

Relocation strategies need not be implemented to be effective. Their goal is to shift risks and redistribute costs outside of the formal boundaries of the firm. The threat of relocation is enough to make significant gains. It is this strategic effort by firms to manage the labor market and governance institutions that characterizes an inter-regional "race to the bottom." However, if firms can elicit cost advantages and concessions from regions through restructuring and redistribution, then the threat of relocation is sufficient. Indeed, actual relocation causes problems for firms given their simultaneous need for firm-specific skills and flexibility. Increasingly, a side effect of the move to implement flexibility is the importance of regional agglomeration economies to firms (Asheim 1992). Firms now need the region. That being said, whether firms threaten to relocate production or slowly reorganize work processes to redistribute the risks and costs of production to the region – the strategic goal is the same: to shape a landscape of production where places bear an increasingly larger share of production costs and firms gain more of the benefits.

The processes and strategies by which firms negotiate agglomeration economies and diseconomies are deeply interlinked with inter-regional competition and intra-regional restructuring. The conceptual framework illustrated in Figures 3.1 and 3.2 describes a two-stage process of competition between and within regions to develop and manage the cycles of agglomeration economies and diseconomies. In Stage 1, the competition is between two regions (Region A and Region B) for an industry specialization. In Stage 2, the competition is between the region and the industry for benefits and costs of economic growth

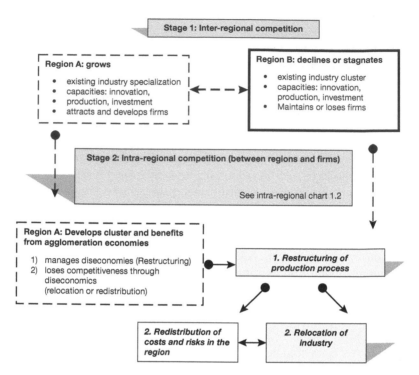

Figure 3.1 Agglomeration diseconomies and regional competition

within the region. Figure 3.1 provides a broad description of the two-stage process of managing growth and change in regional economies. The shaded boxes mark places where firms make strategic decisions to either restructure or relocate production.

In the first stage, there are two regions engaged in competition with each other for an industry specialization. The specialization may be an established industry for which both regions have an existing set of firms – an example would be the competition between Los Angeles and New York City in film and television production. The specialization may also be a new technology – an example is the competition between many US regions for biotechnology firms. In that first stage, both regions pursue a process of investing in applicable technological capacities and institutions and court individual firms. In the end, Region A grows its industry and Region B does not.

Figure 3.2 details the dynamic changes within Region A as agglomeration economies produce growth and increase demand in local markets. The outcomes for Region B are more straightforward. In Region B, firms pursue strategies to manage their decreasing competitive advantage through a restructuring of their production processes. The darker boxes in both figures indicate the strategic decisions made by firms to renegotiate and reorganize the production process

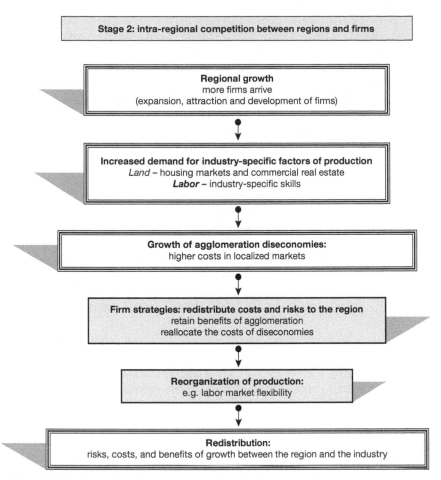

Figure 3.2 Agglomeration diseconomies and regional restructuring

– within the region – in order to manage risks and costs. Labor market flexibility is one strategy that allows firms to retain the benefits of agglomeration economies – a readily accessible skilled workforce – while shifting the costs of reproducing and maintaining that labor market to regional institutions and individual workers. The optics and imaging case in this book provides empirical evidence of this type of restructuring.

Usually these restructuring strategies take a low-road approach – cutting wages and compensation, seeking public subsidies, disinvesting in production facilities and equipment, and pursuing outsourcing and downsizing models. These restructuring efforts generally lead to two outcomes. The first is the relocation of the firm to a region in which the firm sees the potential for developing competitive advantages. The second is a strategy in which the firms manage to displace enough of their production costs to the region to overcome the perceived lack of competitive advantage in that region. The redistribution of risks and costs to the region is a strategy based on re-negotiating regulatory norms and leveraging regional resources. In effect, the region ends up subsidizing production in order to avoid relocation.

For Region A, the growing region, the scenario is somewhat less intuitive than for the declining region. However, the important point is that, for the region, regardless of whether the firms grow or decline, firms will employ similar strategies in order to extract concessions. Figure 3.2 illustrates the processes that take place within Region A as the industry grows. First the industry grows; the region experiences an increasing number of firms and jobs and investments. Second, the demand in regional markets increases. This demand increase is particularly evident in regional labor markets although localized capital markets (e.g. venture capital and small business financing) and housing and land markets will also show increased demand. It is important to note that the resulting price increases in regional markets, as a consequence of agglomeration diseconomies, will not be felt evenly across all markets or sub-markets. Diseconomies in the market for industry-specific specialized skills are likely to appear earlier than shifts in demand for less specialized workers for whom demand may not increase substantially.

Subsequent to the price increases in regional markets illustrated in the third step, firms will make strategic decisions to mitigate the impact of these costs. Here, in managing diseconomies, firm strategies between Region A and Region B converge in terms of the distribution of benefits of industry competitiveness to the region. Region A's firms will make strategic choices to retain the benefits of agglomeration while mitigating the impact of diseconomies. As long as agglomeration economies outweigh diseconomies, firms benefit from remaining in the region.

However, firms are still likely to seek ways to mitigate the rise in costs for the segments of regional markets in which they see increases. A classic mechanism for pursuing this is by increasing the flexibility in employment practices to control wage rates. These strategies include downsizing, outsourcing within the region (creating a two-tiered core and periphery system), and reconfiguring benefits and non-wage compensation packages. With the advent of innovation-based development strategies there is a parallel shifting of research and development capacities to the region in addition to shifting labor reproduction costs. In the final step for Region A, the industry shifts its costs to the region or pursues a relocation strategy when the diseconomies outweigh the scale economies of agglomeration. Figure 3.1 thus shows both Region A and Region B in the same place in the decision-making process – redistribute costs and risks to the region or relocate production. It is not necessary for a set of firms, or even a single firm, to entirely pursue one strategy in its entirety. Most firms make these choices incrementally and reactively. The evidence from the optics and imaging case study shows one example of these incrementally implemented strategies and their effect on one regional market – the labor market – over time.

Conclusions: the role of regional institutions, intermediaries, and governance

Unfortunately, there is growing empirical and theoretical evidence that the causal relationship between agglomeration and growth may be insufficiently understood, particularly in formulating and deploying regional economic development policy (Lovering 1999; Markusen 1999). Indeed, industries and regions stagnate for a wide variety of reasons. Some industries lose their viability in a region because of new competitors, new markets, or new governmental policy.

It is increasingly apparent that regional economic development policy sees agglomeration economies, in terms of a skilled labor market and even high technology infrastructure as a set of necessary but not sufficient elements in a recipe for regional growth. The nature of the agglomeration economies and what sustains them, what undermines them, and what policies can support them, remain significant questions in economic development research (Amin & Thrift 1992; Malmberg *et al.* 1996; Scott 1998).

The "new regionalism," in its theoretical and policy iterations, is based on the positive spillover effects generated by agglomeration: the economic growth and competitiveness, the rising employment, the innovation. However, it is the diseconomies, or the negative spillover effects that are linked, almost inevitably to regional economic decline, firm relocation and job loss, and stagnation. While the economic processes at work in the region are not entirely clear, particularly given that regions do not operate in political or economic isolation from a complex global marketplace, empirical evidence indicates that as demand pushes up costs in local markets (land, labor, or capital), firms seek substitutions. Those substitutions sometimes take the form of relocation, sometimes of production reorganization and outsourcing, sometimes downsizing.

It is primarily networks of institutions that insulate the region from these diseconomies and, to varying extents, mediate the distribution of the benefits of agglomeration between the region and the industry (Peck 1992). Strong labor unions often perform this function, as does strong local governance. Collective bargaining agreements and living-wage ordinances provide predictability of labor costs for firms while recognizing the needs of workers. In the absence of labor unions or local regulation, firms often employ a variety of labor market strategies to buffer themselves from the upward pressure on wages.

Labor market intermediaries, temporary agencies and memberships organizations (to name a few), facilitate the creation of core and periphery systems that limit industry exposure to the risk of wage inflation. These localized systems for mitigating a firm's exposure to risk (primarily the risk of the unpredictability of costs) are not new. Both case studies presented in the following chapters document almost a century of these processes. Distinct institutional forms are clear in most regions and are the subject of many industry and regional case studies (Christopherson & Storper 1989; Stone 1973; Van Jaarsveld 2004; Walkowitz 1978).

Most places, and the politicians and leaders who guide them, have embraced the idea of the competitive advantage model for sustainable, or at least sustained, regional growth. Most places, in order to retain firms and jobs, have compromised their own assets and the governance structures that protect these economic advantages. As industries pit places against each other in a game of spatial competition in which regions are up against other regions, most places no longer can afford to fight to hold the line on community, consumer, and worker protections, ranging from environmental regulation to land use regulation to labor market regulation. Unfortunately, this situation also was described in detail more than 20 years ago when it first became clear that there might be a dark side to flexible production (Harrison 1994a; Holland 1976; Massey 1979, 1984).

Part of the analytical struggle in understanding and illustrating the relationship between firm strategies and regional labor markets emerges from the tendency in research to analyze what is new, what is different, rather than what is perceived as old or mundane. The case studies in the following chapters are distinct not because the industries are new but because they are old. The media industry and the optics industry in the United States have endured in place, in these same regions, for a century. The industries and the cities share an identity, a past, and several possible futures.

Because optics and imaging in Rochester and media in Los Angeles are historical agglomerations, rather than new hubs of innovation or old centers of manufacturing, both industries lend themselves to an analysis of how agglomeration economies evolve over time and the changes in institutional infrastructure and the regional labor market that parallel the process of economic restructuring within firms and industries. Analyses of economic restructuring have typically focused on firms, although there have been some studies on other regional impacts (Markusen 2001). These case studies unpack the process of regional institutional adaptation as both the industry and the region respond to a changing competitive environment.

The time frame only highlights the importance of industry evolution in regional development. As Amy Glasmeier explained in her study of the watch industry spanning nearly three centuries, "My purpose in pursuing such a long industrial history is to investigate the ways in which culture, institutions, organizations, and actors interact with and respond to change inducing events" (Glasmeier 2000).

Firms and regions stay competitive over time because they are able to strategically adapt to a changing environment (Massey 1979). That environment includes a multi-scalar regulatory environment: a global marketplace full of shifting consumer preferences – a world full of technological changes with far-reaching implications. For example, both industries in our case studies are gradually working through the shift to digital photography and digitized production processes. These two industries, at opposite ends of the country, are increasingly less connected by a material product, the film, that once bound them together.

These cases are about heavily concentrated industries, with large firm players influencing market rules at the same time that their sheer size sets regional wage rates (Christopherson 1993). Certainly regional restructuring is more evident in places where the absolute size of the firms makes their efforts to force compromise and concessions more transparent. However, both industries are framed by networks of small firms, many of whom are the creative center of the industry. These networks are full of creative professionals assembled in project teams as consultants and designers, as small partnerships or independent contractors. These labor markets are what the innovation-based economic development strategies seek to model.

Similarly, the embeddedness of these industries in regions means that there is a dense institutional structure in place, mediating the space between the industry and the regions themselves. In Rochester there is an educational and research infrastructure that ensures that optics and imaging technology, and the people who understand it, will be produced in place. In Los Angeles, there is a network of craft-based unions organizing training, certification, industry standards, project teams, and work conditions to facilitate the movement from project to project.

It is the development of these industries within their regions that is at the core of our case studies. The story in these cases is of the restructuring and re-distribution between the industry and the region as well as the relocation of production within and across regions. These stories provide evidence for arguments about the strategic use of scale, including the regional scale, to boost firm bargaining power and control vis-à-vis labor and community. As Jamie Peck pointed out 15 years ago, "the new regime of flexible accumulation seems partial, almost hesitant, in that some colonization of 'new' regions co-exists with a great deal of staying put" (Peck 1992). These cases illustrate exactly how and in what ways that process of staying put plays out.

Section II
Case studies

4 The evolution of the optics and imaging industry

The Rochester restructuring story: changing industries, changing regions

Since the early 1900s, the Rochester region has been the home of large, transnational corporations in two inter-related industries: optics and imaging and photographic equipment and supplies. The large firms in the region are prominent household names including Eastman Kodak, Xerox, Corning, and Bausch and Lomb.[1]

In a city that helped pioneer corporate paternalism, the restructuring that followed the strategic shift from vertical integration to vertical disintegration was a dramatic deviation from a century of job security, internal labor markets, and a predictable employment system (Jacoby 1997). However, even before restructuring reorganized the industry in the 1980s, the industry in Rochester included a large number of small and medium-sized firms (SMEs) tied together through buyer and supplier relations among themselves, the large firms in the region, and a global network of optics and imaging end producers (Sternberg 1992). In recent years this network of small and medium-sized firms has identified itself by technology – photonics – rather than by the traditional end products of the industry – optical, imaging, and photographic equipment. This identification with a shared technology rather than a shared end market underscores the role that small and medium-sized firms play in the transnational supply chain.

While the Rochester region has remained competitive in the global optics, imaging, and photonics industry the contemporary industry has shifted its priorities to match international market conditions (Jacobs 2002). Although Rochester's large firms tend to focus on optics and imaging and consumer and office products, the smaller firms have focused on photonics technologies and a wide array of intermediate markets. In other words, the small and medium sized firms tend not to compete directly with large firms in product markets but rather in the research and development phase of production. Consequently, SMEs are in direct competition with large firms for specialized labor, research and development resources, and intellectual property. The evolution of the SMEs in a more research-driven direction has put them in competition with the regional TNCs for key inputs in the production process (Christopherson & Clark 2000). These inputs are, to a large extent, regionally embedded resources.

As restructuring ushered in new and alternative forms of production organization including small firm networks, flexible production, outsourcing and subcontracting, it also brought in alternative forms of work organization. The emergence of an innovative network of photonics SMEs is paralleled by the reorganization of work for employees throughout the industry. Individual employees found the expected standards of long-term employment, generous benefits, advancement opportunities, and on-the-job training eroding throughout the 1980s and 1990s. The regional labor market, and the institutional infrastructure that framed it, also anticipated the "old employment relationship" (Kapstein 1992; Stone 1981).

Two attributes are considered critical to the ability of co-located firms to become sustainable regional innovation systems and thus create competitive regional economies. The first is cooperation within the network of firms. This cooperation promotes a rapid and flexible response to changing and expanding global markets, and the capacity for innovation. Cooperation among co-located firms enables knowledge spillover from the learning and practice of firms in the co-located network. Knowledge spillover and the "untraded interdependencies" (Storper 1997) produced via a cooperative network essentially make the whole greater than the sum of its parts and lead to a sustainable regional innovation system.

The second attribute is a skilled labor force, which is critical to both innovative capacity and the diffusion of knowledge within and across firms. Agglomeration economies serve as a basis for both the formation of regional labor markets and the organization of firm networks. Agglomeration economies are more than sets of firms defined by end products produced by a set of companies within a fixed geographic boundary (Feldman 2000). Consequently the analysis of industries is intertwined with the institutional assets and infrastructure of the region in which they are located.

The analysis that follows walks through the emergence of the innovative small firm network in Rochester and presents two angles on the consequences of this success for the regional labor market through an analysis of occupational wage rates and the role of labor market intermediaries. The analysis demonstrates that the success of a regional innovation system, measured in terms of small, innovative firms or competitive transnational corporations, fails to take into consideration the uneven distribution of costs and benefits of economic growth to actors within the region.

Regional innovation strategies: small and medium-sized firms

The small firm network of photonics firms in Rochester has long been based in shared technologies and production processes rather than common end products. This was true from the beginning when George Eastman made cameras and Henry Lomb made spectacle lenses. Now, as in the past, firms in the industry make everything from museum lighting to barcode readers, to 3-D to 2-D conversion software to fiber optics to night vision goggles.

Data on employment by firm size in Rochester for the 1990s underscores the emergence of small firms as an increasing factor in regional employment. Small and medium-sized firms under 500 employees added 11,622 jobs between 1995 and 2000 and the number of SMEs increased by 204 firms in the same period (New York State Department of State Division of State Planning, New York State Department of Commerce, and New York State Department of Labor). On the other hand, large firms shed 4,500 jobs but retained the bulk of the area's employment at 253,962 employees. However, the combination of employment in small firms (less than 20 employees) and medium-sized firms (less than 500 employees) exceeds that in large firms both in 1995 and 2000 for the region. The growth of employment in particularly medium-sized firms is consistent with the restructuring trends in the region. How firm size relates to job growth and decline in an era of industry restructuring is an issue of some interest to researchers. Sengenberger *et al.* suggested that, while often the trend toward smaller size firms is attributed to a shift away from manufacturing, as the Rochester case indicates, part of the story involves changes within manufacturing industries themselves (Sengenberger *et al.* 1990).

The literature on firm networks argues that there are three major areas on which agglomeration has a significant impact: innovation, transaction costs, and skills (Storper 1999). The concept of entrepreneurial regions emerges from a long history of attempts to name the characteristics of innovative and successful places. What was once called "industrial atmosphere" was later understood as the "social milieu" (Asheim 1992). A survey of the small and medium-sized firms in the photonics industry addressed these areas.

Recognizing that regional institutions are involved in supporting, defining, and sustaining agglomeration economies, the Rochester case study focused on embedded "institutional infrastructure" and the critical and dynamic role played by the local labor market. While the analysis of these two factors is often minimal in case studies, as compared to the focus on firms, they are nonetheless fundamental sites for investment and regulation.

In researching the regional industry between 2001 and 2004, we conducted a survey of small and medium-sized photonics firms in order to better understand their specific characteristics and their relationships with each other, large firms, and the regional institutional network. We focused on the small firms to examine how they related to markets, and the extent to which they were subcontractors to the major TNCs located in the region (Kodak, Bausch and Lomb, Xerox, Corning). Our survey questions emerged from a series of focus groups and key informant interviews with institutional actors in Rochester engaged in the optics and imaging industry. The focus groups included representatives from firms, universities, civic organizations, trade associations, labor unions, community-based organizations, and public sector agencies. The survey questions were designed to delve into how firms in the industry cooperate and interact with each other in the context of rapidly changing and technologically challenging markets.

The telephone survey was conducted during July and August of 2002.[2] Of the 90 firms constituting the population of optics, imaging, and photonics firms, 57 responded fully to the survey as administered, a 63 percent response rate for the survey.[3] Of these 57 firms, 51.2 percent of the firm representatives interviewed characterized themselves as the owner, CEO, or president of the firm.[4] The firms were primarily small manufacturing firms with 76 percent having between 1 and 50 employees, with none above 500 employees. Seventy-four percent self-identified as manufacturing, rather than design or service firms.

The survey results indicated that the SMEs served widely diversified intermediate markets (see Table 4.1). In addition, although they identified themselves as manufacturing firms, they generally did not manufacture end products themselves but rather optical, imaging, or photonics components for a network of suppliers.

The survey also indicated that 74 percent of firms responding had a past or existing subcontractor or supplier relationship with one or more of the four large, transnational firms in the region – Eastman Kodak, Xerox, Corning, or Bausch and Lomb. The photonics SMEs reported that, 59 percent of them had a present subcontractor/supplier relationship with the "Big Four." In a question aimed at the labor market for industry-specific skills, 42 percent of firms reported that a former or current staff member had previously worked for Corning, Eastman Kodak, Bausch and Lomb, or Xerox.

The survey responses indicated that while most of the small firms were embedded in a regional supplier chain – that is, they interact with one another – their customers were distributed across a broad geographic area (see Figure 4.1).

Table 4.1 Rochester photonics survey: top industries for photonics (SMEs)

Markets for photonics firms in Rochester	Share of firms (%)
Imaging and reproduction	19
Other (aerospace etc.)	16
Semiconductor	13
Scientific instruments	11
Defense	11
Telecommunications equipment	11
Medical devices	8
Consumer products	6
Barcoders/encoders (retail and logistics)	3
Biotechnology	2

Source: Rochester photonics survey, 2002

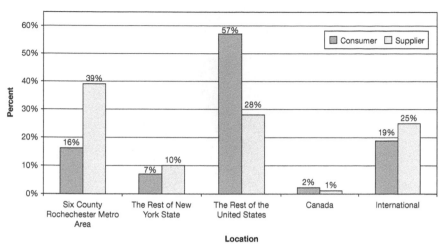

Figure 4.1 Rochester photonics survey: location of customers and suppliers of SMEs
Source: *Survey of Rochester Photonics*, 2002

In this region, SMEs constitute an innovation-oriented capacity as they organize their production towards enabling technologies such as photonics and optoelectronics (see Figure 4.2). Rochester's optics, imaging, and photonics firms serve diverse markets including precision optics, calibration and measuring equipment, medical devices and biotechnology applications, and military and security devices. One firm owner described the industry as an "enabling industry," one that serves product markets based on the wide applicability of the underlying technology. In addition, the product markets are geographically diverse, extending beyond regional and national boundaries. Second, the market for optics and imaging is diverse in terms of end products, serving a broad range of industries.

Because many of the small firms are not making end products for a single market but subcomponents for multiple markets, they are more able to shift research emphasis and production to respond to global market demand. The telecommunications bust, for example, while having a significant impact on Rochester, did not significantly damage the optics and imaging firms that supplied it. The firms shifted their focus to other markets to survive the downturn. This contrasts with photographic equipment, which is largely dependent on consumer expenditures on leisure items such as cameras, film, and movie projectors (Kipnis & Huffstutler 1990). The diversity of markets supplied by Rochester's small and medium-sized optics, imaging, and photonics firms provides the region's producers with insulation against the dramatic fluctuations in any single market (see Table 4.1).

One factor that is taken for granted in the regional innovation literature is the critical role of labor with training in science and engineering. However, our

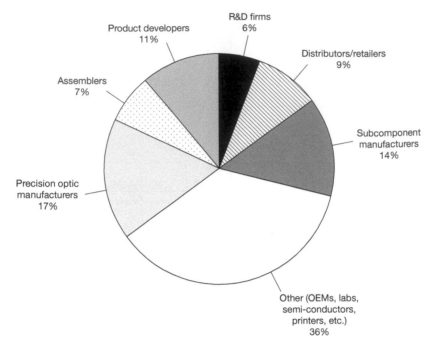

Figure 4.2 Rochester photonics survey: customers of SMEs

Source: Survey of Rochester Photonics, 2002

research indicates that innovative SMEs require a wide range of skilled labor because they are engaged in product commercialization and prototype construction as well as research.

The photonics firms in Rochester were asked two key questions that related to the importance of labor skills: 1) the role labor skills play in their decision to remain in the region and 2) what they identified as key to their ability to grow and expand in the region. Firms identified the quality of the labor supply as the second most important reason for their presence in the Rochester region (see Figure 4.3). When asked the second question: what resources they thought could improve their industry's regional competitiveness, the highest ranked answer was medium-skilled labor (see Figure 4.4).

A clear distinction emerges between the markets in which small and medium sized firms and large firms operate with strong implications for the region. Both groups operate in a global marketplace. However, small firms rarely produce for end-product markets. Instead, large firms dominate end-product markets. As a consequence small and large firms share a global orientation but they rarely find themselves in direct competition in end markets. Indeed, they are in direct competition primarily in input factor markets for skilled labor and research and development capacities within regions.

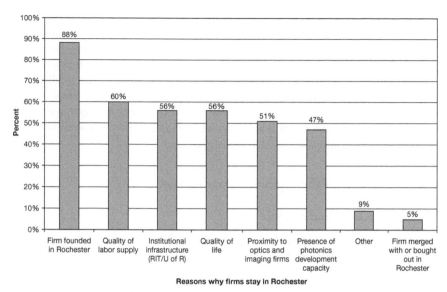

Figure 4.3 Rochester photonics survey: Why do firms stay in Rochester?

Source: Rochester Photonics Survey

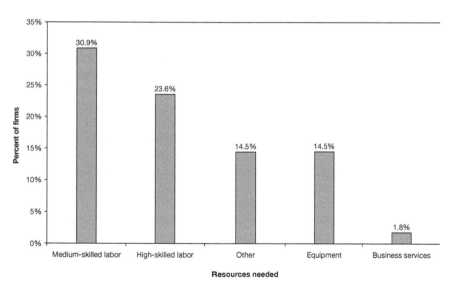

Figure 4.4 Rochester photonics survey: what do firms need?

Source: Rochester Photonics Survey, 2002

This competition in regional markets for labor and research and development resources places small and large firms in an adversarial position (Mudambi & Helper 1998). Both types of firms need the same, regionally embedded, factors of production. Most of these factors are publicly regulated or provisioned, to various degrees. As such, the political and economic power of large firms vis-à-vis the state dominates how public priorities are set. In so far as the small and large firms share the same priorities, there is no disparate impact. However, empirical evidence from the Rochester case and others demonstrates that the small and large firm interests diverge in what they need from research and development institutions and skilled labor markets.

The critical role of the regional labor market

Wage and occupational analysis

While the survey indicated that the SMEs understood that embedded resources produce collective benefits, they viewed them as regional advantages, not inter-dependencies between firms. In other words, in Rochester the firm attachment seems to be to the region and regional resources rather than to a firm network. This recognition of the institutional framework within the region points to the role of tangible institutional actors: labor market institutions, municipalities, universities, nonprofits in building and sustaining firm networks within the region. This stands in contrast to arguments that assert that less tangible assets characterized in a wide variety of terms – knowledge spillovers, entrepreneurial ethos, untraded interdependencies, and innovative milieu – hold the keys to innovation strategies and regional learning (Crevoisier 2004; Feldman 2001).

The next sections add evidence to the argument that a skilled regional labor market is a crucial factor in developing innovative small firm networks and in supporting the evolution of transnational firms in the region. These sections also provide evidence that the success of industry, particularly the success of transnational firms, is largely de-linked from the economic success of workers and residents in the same region. The evidence from the Rochester case indicates that cycles of agglomeration economies and diseconomies and the strategic use of labor market intermediaries shape the distribution of costs and benefits of competitive advantage within the region.

Agglomeration economies function in labor markets in a complex set of competing and contradictory economic processes. Jamie Peck describes the process as follows:

> labor market agglomeration is a contradictory process. While users of flexible labor tend to agglomerate in order to socialize the costs of labor reproduction, in so doing they inadvertently initiate a set of countervailing forces. Agglomeration raises the level of interfirm competition for labor, which may trigger local wage inflation and almost certainly will lead to labor

recruitment and retention difficulties for firms with weak purchase on their labor supply; their hold on secondary labor is inevitably tenuous and, with heightened labor competition, even that tenuous grip may be lost. In this way the agglomeration of secondary sector employers in urban labor markets begins to undermine the utility of such locations. Through overexploitation, the parasites may eventually destroy the host organism.

(Peck 1996)

The spatial clustering of labor transfers the costs of labor reproduction away from individual firms while exposing those same firms to wage inflation through increased competition for labor. The absence of extensive internal labor markets in a flexible production system makes this exposure to the external labor market more risky for firms (Doeringer & Piore 1971).

The benefits of agglomeration economies to firms are substantial, creating cost savings that make individual regions remain attractive sites of production even when competing regions may be able to provide lower cost inputs or assets. Allen Scott outlined the specific agglomeration economies generated by labor market pooling in his articles on the subject of flexible accumulation in the late 1980s and early 1990s (Scott 1988a, 1988b; Storper & Scott 1990). In short, a significant regional pool of both workers and jobs: 1) reduces labor turnover rates; 2) reduces the costs of turnover for workers by providing more opportunities; 3) reduces job search costs by making information gathering easier; 4) builds on a tendency of workers in secondary labor markets and firms that employ secondary workers to co-locate; and 5) contributes to a recursive process, over time, between work conditions and worker expectations thus creating a space that matches with the industrial structure (Peck 1992). The end result is a mutually beneficial arrangement for industry, workers, and the region.

The optics and imaging industry in Rochester endures, in situ, in significant measure because of the strategic use of the region by the dominant firms in the industry (Christopherson & Clark 2000; Clark 2004; Drennan 1998; Pendall *et al.* 2004). The large firm corporate culture was so embedded in the city that its civil rights movement in the 1960s centered on the discrimination in the hiring at Kodak and Xerox. Also, commercial activity in the city was closely linked to firm practices. For example, when Kodak distributed the annual bonus checks, Rochester department stores conducted "St. Kodak's Day" sales (Sethi 1970; Wadhwani 1997). Since the 1980s, the large corporations in the region have pursued downsizing and outsourcing strategies. While overall regional employment has remained consistent, per capita income has steadily declined. In Rochester this is particularly dramatic because income in the region not only declined, but it did so relative to the state and the nation (see Figure 4.5). Rochester had long been known as a region with a high standard of living, attributed in part to the standard of corporate paternalism set by its large employers (Jacoby 1997).

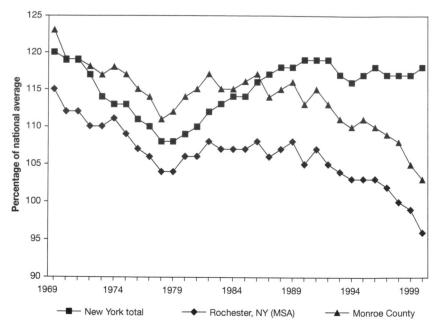

Figure 4.5 Per capita personal income as a percent of the US average (1969-2000)

Source: Detailed County Annual Tables of Income and Employment by SIC Industry, 1969-2001 (Ca30-Ca45), 1969-2000 (Ca05 and Ca25) Local Area Personal Income Statistics, Regional Economic Accounts, Bureau of Economic Analysis, U.S. Department of Commerce, 2002 [cited January 2003]. http://www.bea.doc.gov/bea/regional/reis/

An analysis of occupational wage rates in the optics and imaging industry indicates that Rochester lost ground in its historic role as a high wage region in spite of its ongoing innovation in photonics. National wage rates for several industry-specific occupations are higher than they are in the Rochester region. This evidence was confirmed in interviews with small firms and economic developers who pointed to the difficulty in attracting skilled labor to the region largely because of the lower compensation.

Specifics of the occupational analysis in the Rochester case are provided in Table 4.2 through Table 4.4. Tables include the occupations with the highest occupational location quotients organized by major occupational categories. Tables also include wage data and a "relative wage," which indicates whether the occupations in the region were compensated (on average) above or below the national average. Because optics, imaging, and photonics is an industry with both a research and development component and a significant precision production component, selections from these broad occupational categories are included (Kipnis & Huffstutler 1990). Table 4.2 covers production occupations.

All of the occupational categories in the tables exceed 1.2 for their location quotient. Of these production occupations, which totaled 65,380 jobs in the year 2000 for the Rochester metropolitan statistical area, the highest location quotients (in excess of 7) appeared in a division of machine tool setters and food cooking machine operators. The location quotient for inspectors and testers was over two. Both machine tool setters and inspectors and testers are highly correlated with the optics and imaging industry. Emphasizing the role of precision production, the computer-controlled machine operator category also had a high location quotient although wage levels remain lower than the national average. High location quotients were also associated with coating occupations, which is consistent with innovations in thin film coatings and other lens and optical coatings.

Many of the occupations with high location quotients did not have higher wages than the national average. In other words, occupations in the dominant regional industry with the highest number of workers had lower wages than in less competitive regions. This is particularly notable because optics, imaging, and photonics is an innovation-based industry with an historically well-developed small firm network. Indeed, for some occupations, the wages were considerably lower. For other popular jobs like machinists and tool and die makers, the wages were also below the national mean despite their role as supplier occupations to optics and imaging.

The larger number of workers in these occupational categories results in a regional pool of workers with industry-specific skills. Thus firms face less pressure to bid up wages because potential employees are in ready supply. This downward wage pressure provides another motivation for firms to remain in regions with industrial specializations because a large pool of skilled employees minimizes labor costs (Hansen *et al.* 2003; Peck 1996; Pendall & Christopherson 2004). This is consistent with the literature on agglomeration economies and labor markets although it differs from arguments about regional innovation strategies and the benefits of high-tech agglomerations (Drennan 2002; Krugman 1991). The occupational data from the Rochester case raises an important question about the balance between wages, skills, and innovation.

Table 4.3 displays architecture and engineering occupations with the highest regional concentrations. This table focuses on the "knowledge-based" occupations within an "innovation-based" industry (Florida 2002a, 2002b). Rochester firms and educational institutions have long focused on research and development and there is a strong engineering emphasis in the region. The region includes two research universities specializing in engineering professions, the Rochester Institute of Technology and the University of Rochester.

For this category of occupations – engineering – the wage pattern shifts from that of the production and precision production occupations. The wages were almost all higher for these occupations, with the exception of architects and electronics engineers. Materials scientists made almost $12 more per hour than the national average while materials engineers made $8 more. The higher

Table 4.2 Production occupations, location quotients for employment and wages for the Rochester MSA, 2000

Occupation title	MSA employment	National employment	MSA mean hourly	National mean hourly	Employment location quotient	Relative wage
Production occupations	65,380	12,400,080	$13.71	$12.72	1.28	1.10
First-line supervisors/managers of production and operating workers	4,450	769540	$22.45	$20.68	1.41	1.11
Electrical and electronic equipment assemblers	3,330	367150	$9.75	$11.03	2.20	0.90
Electromechanical equipment assemblers	520	72550	$11.08	$11.81	1.74	0.96
Team assemblers	6,930	1306430	$11.62	$11.29	1.29	1.05
Food batchmakers	380	67320	$9.54	$10.71	1.37	0.91
Food cooking machine operators and tenders	1,040	36020	$7.27	$10.49	7.02	0.71
Computer-controlled machine tool operators, metal and plastic	1,670	162360	$12.28	$13.84	2.50	0.91
Forging machine setters, operators, and tenders, metal and plastic	330	53950	$12.12	$13.30	1.49	0.93
Lathe and turning machine tool setters, operators, and tenders, metal and plastic	2,620	84020	$15.96	$14.27	7.58	1.14
Milling and planning machine setters, operators, and tenders, metal and plastic	410	35610	$9.97	$14.00	2.80	0.73
Machinists	3,020	420320	$14.42	$15.20	1.75	0.97
Model makers, metal and plastic	190	10540	$21.71	$17.10	4.38	1.30
Patternmakers, metal and plastic	70	8290	$16.52	$15.88	2.05	1.06
Molding, coremaking, and casting machine setters, operators, and tenders, metal and plastic	1,630	158280	$13.25	$11.36	2.50	1.19
Multiple machine tool setters, operators, and tenders, metal and plastic	830	109950	$12.44	$14.11	1.83	0.90
Tool and die makers	1,550	131080	$18.99	$20.07	2.87	0.97

Occupation						
Plating and coating machine setters, operators, and tenders, metal and plastic	290	54760	$13.19	$11.82	1.29	1.14
Prepress technicians and workers	550	104920	$15.87	$15.31	1.27	1.06
Stationary engineers and boiler operators	360	56330	$20.21	$19.94	1.55	1.03
Separating, filtering, clarifying, precipitating, and still machine setters, operators, and tenders	190	36110	$13.39	$13.77	1.28	0.99
Crushing, grinding, and polishing machine setters, operators, and tenders	260	45010	$12.52	$12.60	1.40	1.01
Mixing and blending machine setters, operators, and tenders	790	111480	$17.36	$13.05	1.72	1.36
Cutting and slicing machine setters, operators, and tenders	1,260	82450	$16.52	$11.98	3.71	1.41
Inspectors, testers, sorters, samplers, and weighers	4,830	571220	$15.22	$13.47	2.06	1.15
Coating, painting, and spraying machine setters, operators, and tenders	2,060	103650	$17.42	$12.09	4.83	1.47
Paper goods machine setters, operators, and tenders	1,710	121300	$13.07	$13.32	3.43	1.00

Source: Rochester MSA, 2000 Occupational Employment Statistics (OES) Survey 2000: Regional and National Comparison of Employment by Occupation.

Table 4.3 Architecture and engineering occupations, location quotients of employment and wages for the Rochester MSA, 2000

Occupation title	MSA employment	National employment	MSA mean hourly	National mean hourly	Employment location quotient	Relative wage
Architecture and engineering occupations	15,350	2,575,620.00	$27.93	$25.99	1.45	1.10
Architects, except landscape and naval	560	74,390.00	$25.11	$26.93	1.83	0.95
Electronics engineers, except computer	670	123,690.00	$30.83	$31.97	1.32	0.98
Environmental engineers	2,100	48,270.00	$32.66	$28.70	10.58	1.16
Materials engineers	210	24,430.00	$37.12	$29.05	2.09	1.30
Electrical and electronics drafters	240	38,470.00	$19.85	$19.43	1.52	1.04
Electrical and electronic engineering technicians	1,390	244,570.00	$21.01	$19.81	1.38	1.08
Electro-mechanical technicians	310	40,770.00	$19.20	$18.57	1.85	1.06
Industrial engineering technicians	870	65,220.00	$23.89	$21.31	3.24	1.14
Materials scientists	160	8,660.00	$42.04	$30.28	4.49	1.42

Source: Rochester MSA, 2000 (Occupational Employment Statistics (OES) Survey 2000: Regional and National Comparison of Employment Occupation).

wages are likely indications of a higher demand for engineering professionals. Interviews with firms in Rochester revealed that the use of independent contractors for engineering is increasingly common. The higher wages may reflect the difference between independent contracting rates and the wages paid to employees as a compensation package including health benefits and pensions (Stone 2004; Van Jaarsveld 2002).

In addition to a demand for engineers, interviews consistently revealed a shortage of computer professionals. Firms repeatedly reported difficulty in filling these computer-based positions. While Rochester had a high location quotient for research computer scientists, the general category of computer occupations had wages below the national average. Further, the salaries for computer related occupations with high location quotients were in fact lower than the national average (see Table 4.4). Again there is a question of what sort of employment relationship (permanent full-time vs. part-time and/or contingent) is typical and, whether the same phenomenon occurring with the concentration of precision production workers is affecting wages here with computer professionals. However, the occupational evidence clearly raises questions about whether an innovation-based industry brings with it higher wages for all workers or simply those at the top of the ladder.

While lower wages are never good news for workers, they do benefit firms who can hire skilled labor at relatively lower costs. Again, the strategic management of the processes of restructuring and redistribution within the region result in a divergence between the interests of workers and firms. What is good for the industry is not always good for the region, even when the industry remains in place and the region retains production. The question may come down to whether the region is identified with its workers or its firms.

Labor market intermediaries and institutions

A second lens through which to analyze the impacts of success in the global economy of the workers in the region is through outsourcing and subcontracting and the role played by private, for-profit labor market intermediaries in the regional economy. A Labor Day special segment on PowerLunch, CNBC's financial news program, featured an analysis of the "new trend" of outsourcing work overseas. PowerLunch focused on some of the largest U.S companies, namely GE and Microsoft, shifting information technology work to India. The analysts and firm consultants predicted the trend would increase over time for two reasons. First, the skills of international workers were sufficient to perform the work and second, the labor costs savings were substantial.

What PowerLunch failed to mention was that this trend is not new, but rather another iteration of a series of labor strategies meant to implement "flexible employment practices" deployed by US firms for the past quarter century (Bluestone & Harrison 1982; Cappelli 1997; Stone 2004). One such strategy, less detailed in today's popular business media, involves the subcontracting for

Table 4.4 Computer and mathematical occupations, location quotients for the Rochester MSA, 2000

Occupation title	MSA employment	National employment	MSA mean hourly	National mean hourly	Employment location quotient	Relative wage
Computer and mathematical occupations	11,350	2,932,810	$27.98	27.91	0.94	1.02
Computer and information scientists, research	380	25,800	$33.29	35.3	3.58	0.96
Computer software engineers, systems software	1,930	264,610	$32.33	34.08	1.77	0.97
Network and computer systems administrators	1,240	234,040	$23.75	25.81	1.29	0.94

Source: U.S. Bureau of Labor Statistics, Rochester MSA, 2000 Occupational Employment Statistics (OES) Survey 2000: Regional and National Comparison of Employment by Occupation.

labor within regions rather than overseas. The familiar term "temp firm" has become common parlance for the array of third-party labor subcontractors who deploy their own employees to client firms on a contractual basis (Van Jaarsveld 2002). This outsourcing, while perhaps less visible than the movement of computer programming jobs to India, has become common within industries and regions (Lautsch 2003).

In the past, firms often outsourced production during cyclical periods of market decline. In the "new" employment structure restructuring occurs, not cyclically, but constantly (Cappelli 1997; Harrison 1994a). Firms often retain employees in their "core competencies" while outsourcing other functions, although job tenure and stability seems to be decreasing for both groups (Stone 2001). However, particularly for cities and regions with dominant industries and significant agglomerations, outsourcing often means that firms subcontract for workers from the same pool of skilled labor from which they once hired directly. In some cases, they even indirectly rehire the same personnel through a third party contractor (Florida 2002c). For firms and industries requiring a high degree of technical expertise in production or with a heavy reliance on research, development, and design, the externalization of workers from the firm brings both risks and rewards. The question in this case is whether and to what extent the agglomeration, or spatial clustering, of firms facilitates the outsourcing of specialized labor within a sub-national region (usually a metropolitan statistical area corresponding to a labor market)? This research also asked how firms, labor contractors and client firms in the industry: 1) insulated themselves from the risks (shortage of skilled workers, competitive wage pressures, intellectual property loss) of externalization; and 2) maximized the rewards associated with lower labor costs.

The optics and imaging specialization of the Rochester labor market creates a positive feedback within the region as both a result of industry agglomeration and a significant benefit to firms. This process encourages firms to remain in the region and benefit from the economies of scope and scale that agglomeration offers (Drennan 2002). However, as several researchers have argued, the agglomeration of workers with specialized skills, over time, places upward pressure on wages, thus mitigating these cost benefits (Harrison 1994a; Peck 1992, 1996). In Rochester the internal labor market structure protected large firms from these external market pressures and in some occupations it continues to do so (Befort 2003). For other occupations the internal labor market has broken down in favor of more flexible employment relationships (Christopherson & Clark 2000). These occupations are thus exposed to external market forces and subject to this predicted "bidding up" of wages in times of low unemployment and high demand. In Rochester this has not happened (Pendall & Christopherson, 2004; Pendall *et al.* 2004). One explanation may lie in the dense network of private labor market intermediaries that facilitate contingent employment and placement in the optics and imaging firms.

In Rochester, a city historically dominated by large manufacturing employers, the issues of labor relations and labor control have long been a central question in production organization. New York State has traditionally been a stronghold for organized labor, with the percentage of unionized workers at 25 percent even as the average for the country hovers around 13 percent.[5] Rochester, however, was the exception to the rule for New York cities, mirroring national, rather than state, trends. Rochester's largest employers – principally Eastman Kodak and Bausch and Lomb – resisted organized labor throughout the twentieth century using scientific management techniques and corporate welfare to provide marginally better working conditions in an effort to instill firm loyalty (Clark 2004; Jacoby 1997). While Xerox's manufacturing workforce is unionized, many manufacturing firms in the optics and imaging sector have followed the Kodak and Bausch and Lomb lead which included the use of internal labor markets, seniority systems, higher wages, benefits and bonuses, on-the-job training, and predictable promotion to discourage organizing and regularize the labor market. This model has dominated labor relations in Rochester until recently and mitigated the need for external labor market intermediaries (external to the firms themselves) such as organized labor, public employment and training networks, and private intermediaries, like temp firms (Doeringer & Piore 1971).

The shift in the mid-1980s away from this internal labor market model and towards a flexible employment system changed the character of employment relations in Rochester. The outsourcing trends associated with vertical disintegration had a disproportionate impact on Rochester because the corporate welfare model of labor relations was so dominant in the regional labor market (Jacoby 1997). As the steady, relatively high-paying jobs in medium-skilled manufacturing occupations became outsourced and downsized, the labor market was exposed to wage and productivity pressures that it had been isolated from by internal labor markets (McKelvey 1973).

With the change in employment practices by Rochester's largest firms flowed the need for labor market intermediaries to fill the gaps in the labor reproduction process that had once been addressed within the firm. As firms externalized labor costs, the burden of education and training, recruitment and placement, career ladders, and even certain benefits (health, child care, etc.) became unmet regional needs or fell under the jurisdiction of non-profit and public agencies. The significant expansion of optics and imaging programs at the Rochester Institute of Technology and the University of Rochester in the mid-1980s, with public investment in excess of 20 million dollars, is but one example of the shifting responsibility for education and training capacity and an economic development focus on research and development (Lovering 2001; Sternberg 1992). While the incremental shift of these costs has been subtle, over the past 25 years they have had a cumulative impact. The large regional firms have reset the standard employment relationship from corporate paternalism to a core and periphery, lean and mean model (Harrison 1994a).

While employers in Rochester have thus far avoided many education and training shortages due, in part, to the steady stream of downsized employees from the "Big Four" who have industry-specific job skills, the role for a job placement labor market intermediary remains significant on both the supply and demand side of the labor market (Christopherson & Clark 2000). While education and training institutions have been one of the strengths of the Rochester economy throughout the twentieth century, it is the production jobs, not the engineering or research jobs, which have been most profoundly disrupted by the dismantling of the internal labor market and outsourcing of production. Into this gap a network of private, for-profit labor market intermediaries (LMIs) have moved to provide flexible employment for firms throughout the region and provide placement services for workers. Evidence indicates that private LMIs serve as an implementation mechanism for the wage control and flexible production system emerging in Rochester (Peck & Theodore 2001).[6]

Temporary employment agencies have come a long way from the 1960s and the image of "Kelly girls" and typing pools. The Kelly girls now show up as the managers and presidents of the small local temp agencies and multi-national branch locations of these twenty-first century versions of their former workplace.[7] In Rochester the temporary agencies are keenly aware of the evolution of the industry from on-site typing pools to outsourced plant floors.

In the literature concerning flexible employment practices, research into the role and function of contingent work and temporary firms is a topic of some concern as the number of contingent workers has increased significantly throughout the 1990s, as the entire industry expands beyond temporary secretaries and into manufacturing and the high-tech labor market (Stone 2001, 2004; Van Jaarsveld 2002).

During the course of this research in Rochester, the US national economy went into a recession that had a significant impact on the telecommunications industry.[8] The manufacture of telecommunications components is one of several applications derived from optics and imaging and thus that downturn had an impact on employment patterns in Rochester. The temporary employment firms in this study were able to talk about their role in the labor market both in terms of expansions and recessions. The trends in contingent work in this case thus show more variation than the persistent upward trend seen in the national data.

For example, several temporary firms in Rochester pointed out that, because they are "just-in-time" labor subcontractors, they are the "canary in the mine" for changes in labor demand.[9] Certainly this informal understanding of the role of temporary firms in the regional labor market is supported by the national, quarterly Manpower survey which assesses the demand (and projected) demand for temporary employees nationwide and is frequently used by analysts as an indicator of general employment trends (Manpower Inc. 2003).

Table 4.5 indicates the trends in employment and wages for the temporary employment agencies in the counties located in the Rochester region.[10] While

Table 4.5 Statistics of temporary employment for counties in the Rochester region, New York, 1997-2003

	1997	1998	1999	2000	2001	2002	2003
Monroe County							
Employment	9,006	9,891	11,506	11,496	9,639	8,939	8,615
Number of Firms	101	115	135	137	142	144	130
Weekly pay ($)	381	372	365	347	420	428	460
Annual pay ($)	19,882	19,343	19,002	18,041	21,861	22,254	23,946
Wayne County							
Employment	959	1,212	1,837	1,882	592	281	275
Number of Firms	9	9	9	8	9	10	8
Weekly pay ($)	290	285	314	317	421	571	409
Annual pay ($)	15,056	14,810	16,335	16,468	21,876	29,672	21,251
Ontario County							
Employment		617	486	405	419	434	689
Number Firms		10	9	9	11	11	9
Weekly pay ($)		303	276	283	311	326	365
Annual pay ($)		15,772	14,364	14,740	16,151	16,961	18,956
Genesee County							
Employment		474	532	643	643	666	614
Number of Firms		3	3	3	3	4	4
Weekly pay ($)		218	238	245	254	267	290
Annual pay ($)		11,349	12,402	12,718	13,231	13,878	15,065

Source: U.S. Bureau of Labor Statistics, 2000

there are significant problems with how and whether contingent workers are accurately counted, the data does lead to some general observations for the regional labor market. First, the trend of increasing employment in the contingent sector, particularly in the central county of the region, Monroe County, shows a precipitous drop between 2000 and 2003, presumably in response to the decrease in demand associated with the recession. Wages however, increase.

Further, although employment declines, there does not seem to be a corollary significant decrease in the number of firms. One could surmise that because temporary firms make their money from the time each worker works, the increase in the wages would offset the firms' (potential) losses from a decrease in workers. Intuitively one might hypothesize that workers' wages would decrease as employment demand decreased but in fact wages increased. This was presumably for two reasons: 1) workers, particularly skilled workers, may be working more hours as the number of total workers deceases; 2) work available for lower paid workers is often the first work to evaporate so the average wages for temp firms may thus increase. Perhaps what the chart underscores is the complex way in which contingent work and private, for-profit labor market intermediaries function within the broader labor market. In many ways these temporary agencies distort the traditional labor market by making both labor supply and labor demand more opaque.

Labor markets, particularly those tending toward flexible employment models and practices, are geographically fixed. Although establishing the definitive boundaries of regional labor markets is widely considered a nearly impossible task, it is broadly accepted that these boundaries, within the elasticity of transportation changes and residential options, do exist even if they cannot be mapped with precision (Peck 1996; Peterson & Vroman 1992). The fixed nature of regional labor markets gives them, and the labor market intermediaries that facilitate job placement, training, advocacy, and other services, distinct and localized characteristics. These characteristics, and the norms and customs that determine regional employment relations, are often shaped and reshaped by the strategies of large regional employers who function as dominant regional actors within the labor market (Grantham & MacKinnon 1994).

The temporary employment firms in Rochester indicated in interviews that the role of Xerox, Eastman Kodak, Bausch and Lomb, and Corning in the regional labor market affected them directly and affected the labor outcomes for smaller firms in the region. The wages and "mark ups" in firms that the temporary agencies saw as subcontractors were mentioned as problematic by several agencies:

> That poses a challenge in the fact that because they're then selling their product to Xerox, there's an added layer on top from a cost perspective so the smaller firms don't pay as much obviously as the bigger ones and our mark ups usually can be bigger in the smaller firms but it's harder to recruit because the wages are lower. It's easier for us to place people obviously at a higher wage job but the result to us as far as an hourly billing rate usually is the same but at Kodak or Xerox, they're getting, the employee gets the bigger percentage of that hourly billing rate and at the smaller firm, you know they're making less money.[11]

These comments from an Adecco manager indicate that while firms, large and small, are using temporary employees, the larger firms have managed to maintain their dominance as the "employer of choice" for qualified employees without even actually being their employer. Again, the regional dominance of the large firms within the labor market is passed on through the pay structure of the temporary agencies. Large firms get better employees but no longer have to pay a premium for the privilege. In fact, that cost appears to be absorbed by the temporary agency as well as the legal risks associated with being a direct employer. Further the productivity losses associated with using less-skilled employees are passed on to smaller firms and subcontractors.

The temporary agencies identify themselves as labor subcontractors and their mark up as the way they charge for the service they provide: recruitment, placement, assessment, and screening. All of the firms interviewed in the Rochester area saw both pay rates and mark up rates as heavily determined by the large employers in the region, principally Xerox and Kodak:

> Kodak and Xerox really drive the pricing in Rochester for most of our customers, whether it's manufacturing or clerical because both of them have these extensive cost-reduction programs year over year where they require all vendors, whether it's me or whether it's, you know, Accucoat to give them double digit productivity gains every single year. You have to show that you've reduced your overall cost by 10%, whether it's price, whether it's in your profit fees, whether it's in your services, somehow you have to be able to prove and demonstrate to Kodak and Xerox that you have saved them double digits every single year which is hard to do. It's extremely hard to do . . . It's a lot easier for a manufacturing company to do that . . . but when I deal with only people and my margins are only this big to begin with, unless I affect their wage which we've not been in a position in our labor market to affect wages in the last 5 years . . . you know, 80% of my dollar that I charge Kodak is made up of what the employee gets . . . We're finding that the smaller companies are now starting to ask us for double digit productivity because they have to show Kodak or Xerox that they got it . . . I sat in a meeting last week with a very small customer he said my expectation is you will give us double digit . . . and he said oh, do you know that program. And I said I know it a whole lot better than I'd like to so the market has absolutely been impacted by those types of programs run by the big customers.[12]

The Adecco manager went on to say that this "double digit productivity" program is not unique to Rochester but rather is a corporate strategy that Adecco, as an international firm, encounters across regional labor markets dominated by large employers with subcontractor networks. The productivity policies of large firms throughout their subcontractor networks have an especially problematic effect on their labor suppliers. The productivity policies do not always directly set wages at a specific rate but rather create pressure on both the temp agencies and their other subcontractors to keep wages low and prevent wage increases over time (Phelps 1997).[13] In this way they keep their own wages relatively higher than their labor competition and prevent them, and the temp firms, from allowing increases in regional wages. Other temporary agencies also mentioned the pressure client companies put on intermediaries to keep costs and wages low.[14]

Perhaps the most vivid telling of the wage pressures put on intermediaries and the regional labor market came from the owner of the Rochester franchise of Manpower, Inc.:

> well of course Kodak was calling the shots in terms of setting wage salaries. But a change occurred in wage setting through the use of contract and

staffing services. I'll give you an example from three or four years ago. Xerox was bidding a contract. Xerox would employ 2,000 temps of one kind or another a year in Monroe County. Big numbers. They come to us and say look guys, we need to cut our costs . . . We're going to allow you to pay a maximum bill rate and we're going to tell you what it is. We're not even going to let you bid on it. If you want to participate you come in and do it and, oh by the way, it's less than last year . . . we don't think you're going to have a hard time finding people to fill these jobs because we are going to supply them to you. Well that often happens, yes a former Xerox employee goes back in through a temporary service. You mean they were making $14 an hour, you expect us to pay them $10 to do the same job.[15]

Manpower managers pointed out that, like Adecco, they were given these ultimatums throughout the country by large regional employers, including General Motors, Ford, and IBM. Further, the referral of laid-off employees to a temporary agency, particularly if the former employer knows that temporary agency has openings, reduces the former employer's risk of paying unemployment insurance while their laid-off worker looks for a new job. This process of laying off workers while referring them to a temporary agency with a preexisting order – from that same company – was reiterated as "normal" by a manager at Burns Personnel.[16] The manager at Burns Personnel also bemoaned the wage pressures exerted on intermediaries by the large regional employers through double digit productivity requirements and "reverse bidding" for contracts.[17]

The issue of liability of the employer, under existing employment laws and other regulations, seems to be a critical one as corporations attempt to limit exposure to labor regulations such as worker's compensation, unemployment insurance, and other liabilities. Firms attempt to pass that exposure and that risk on to the temporary firms:

> We're being squeezed totally because companies like a Kodak or a Xerox think they know what your costs are because they have certain costs built into their own hiring schedule . . . So they think they know, but they in many cases are the ones who create havoc that the smaller companies have to pay higher premiums on certain things because of their exposure on certain things like unemployment or worker's comp. If I get two people hurt on workman's comp and they have big claims, next thing you know I'm paying thousands of dollars for workman's comp insurance. My exposure – everything is my exposure.[18]

Notably, the labor subcontractors for the large firms in Rochester negotiate contracts with the purchasing departments rather than human resources which is more typical for smaller firms. Many of the temporary firms have multiple sites in the region in addition to providing on-site management for larger contracts. Adecco for example, has 12 locations in the greater Rochester area in stark contrast to the single downtown location for the public sector's "One-Stop" employment services center.

According to long-time Rochester temporary agencies, Kodak did not begin to use temps until the mid-1960s and even then temps were not allowed "on-site" but worked in a separate facility under "sole-source" contracts.[19] More recently Kodak, and most other firms, have used multiple vendors, with the double-digit productivity increases a requirement. As a result, some of the smaller, local temp firms, unable to mitigate losses through other contracts, have abandoned working directly with Kodak and now work as second tier subcontractors to the larger employment agencies. Some large firms prefer to retain the direct relationship with a "sole source" provider, for example Bausch and Lomb only uses Kelly Services.[20]

The issues surrounding second tier subcontracting are interesting in that the temporary agencies who compete with each other both for workers and for employer contracts also collaborate in order to manage large orders, a characteristics often attributed to firms in industrial districts (Benner 2003; Markusen 1996). As the Industrial Management Council describes it:

> We operate a little bit differently than your usual temp agency. We have a lot of member companies that will only use our services and we often can't fill all of their orders. So we have subcontracting arrangements with a lot of our competitors. In turn, we don't have the volume to staff a Kodak or some of the larger firms . . . so we do supply temporary employees to places like Kodak and Xerox, but it would be through Adecco or through Burns. They may get an order for 50 people and they might not have 50, so we'll give them 10 of ours. And they do the same thing for us. It's a very unique working relationship that has developed here in Rochester.[21]

The problem of filling large volume contracts was particularly acute before the recession and telecommunications crisis. Manpower reported in 2000 that they rarely filled more than 50 percent of their skilled orders daily and 70 percent of unskilled orders.[22] In an environment with such a tight labor market the motivation for the collaboration amongst competitors, described by the manager quoted above, becomes clearer. In addition, the extraordinary efforts to use intermediaries to manage wage rates make more sense. This issue of labor market control and wage pressure filters down to small, local employment firms as well as the multi-national temp agencies.

In this regard, temporary employment agencies, as "just-in-time" labor suppliers, must respond directly to client firms. Further, because the majority of temporary employment firms base profits on volume, they are particularly captive to the labor market strategies of large employers. Thus the establishment of norms in the flexible employment system is not just a result of the direct employment strategies of large firms but is also mediated and magnified by temporary employment firms, their labor supply, and management strategies (Krugman 1991). These norms extend to wages and work to suppress wage inflation (Van Jaarsveld 2000).

It is striking how the temporary firms in Rochester viewed their role in the labor market not as price setters for occupations but as price takers. In a traditional framework this would indicate that the employee was setting the wage rate but, in fact, in this context it means that the client firm is setting a bill rate. Thus the wage rate continues to be an outcome of the client firms' willingness to pay for labor. Further, the mark-up rate of the temporary firms in Rochester remains around 30 percent, with one temporary agency indicating that without additional services – extensive background screening, skill diagnostic assessments, extensive personnel files – they were happy to operate at 28 percent indefinitely.[23]

It is the additional human resources functions, added by temp firms because of client firm demand and the prospect of higher bill rates, which create real problems for other labor market intermediaries. There remain, however, many traditional intermediary functions that temp firms find themselves unable to engage in because of the flexible nature of their employment patterns. Primary among these is the issue of worker training.

The increasing use of temporary employment agencies is a firm strategy with significant implications for the regional labor market. The decline in firm reliance on internal labor markets has led to an increased dependence on external labor markets, or the regional pool of labor outside of the firm. This shift from internal to external labor markets for significant portions of the workforce of large and medium-sized firms is thought to increase the production flexibility. Generally speaking, even before the downsizing characteristic of the 1980s and 1990s, American firms were more inclined to view labor as a "variable cost and not an asset" (Barnes & Gertler 1999). The attempt then to externalize labor costs, in order to mitigate those variable costs through firm labor market strategies that emphasize flexibility, is by no means surprising.

Unfortunately the move towards increasingly flexible labor market strategies exacerbates some of the endemic problems in the US labor market: namely, the problems of skill formation and wage stagnation, including a lack of basic benefits (Christopherson 1989; Peck 1992). While the internal labor market addressed these issues by on-going training, seniority, career ladders and mentorship, the flexible employment systems leaves these systematic problems to the public sector and individuals, broadly construed as the region.

This case study demonstrates how temporary employment firms organize themselves individually and collectively in an environment in which a few large institutional actors (dominant employers) set the wage and profit parameters. The Rochester case illustrates how temporary firms function as labor market intermediaries and in particular how these labor subcontracting firms: 1) are captured by high volume client firms and serve to amplify the employment norms and customs preferred by high volume regional employers throughout the labor market; and 2) undermine the efficacy of other LMIs (both their for-profit competitors and public and nonprofit LMIs) and influence the type of human resource services they provide.

The lack of effective and influential public sector labor market intermediaries in the US is peculiar to the US as compared to other industrialized economies. This gap is underscored in case studies involving labor market intermediaries in Germany and Italy and to a lesser extent Canada and the United Kingdom (Christopherson 1993, 2002; Katz & Darbishire 1999). The dominance in Rochester of private, for-profit labor market intermediaries in the process of recruiting, assessing, and placing workers creates a significant challenge for organized labor, the public sector, and nonprofit organizations whose view of labor extends beyond a simple commodity. The dominance of private intermediaries, and the lack of effective engagement in workforce development by employers and the public sector sets in motion a looming skills crisis which has the potential to undermine the successes of a once specialized labor market.

Adaptation in the region: lock-in or resource exploitation

The symbiotic relationship between the regional labor market and firm networks produces a constantly negotiated process that redefines boundaries, costs, benefits, risks, and regulations. This continuous restructuring process is also a spatial process in two ways. First, the intra-regional effort to rescale away from the city to the region reshapes the sphere of regulation and responsibility with profound political economic consequences. And second, the inter-regional effort to force regions into competition with each other, whether real or perceived, reinforces a concessionary logic that undermines some of the most basic and fundamental institutional foundations of successful agglomeration economies. In effect, the readiness of regions to concede assets to firms compromises the positive regional impacts of agglomeration economies. Firms strategically use inter-regional and inter-jurisdictional competition to extract the benefits of agglomeration economies from regions rather than share the benefits with regional residents.

In the Rochester case, the concessions have meant the slow erosion of wage rates and disinvestments in the city and thus in the cultural and civic amenities that constitute a high quality of life. It is not a question of whether the region should adapt to global competition and industrial change, clearly that adapt-ability and the ability to stave off agglomeration diseconomies has assisted in the retention of the industry in the region. However, it is not clear how and in what ways the region could have or should have renegotiated the distribution of risks and altered the regulatory structure to maintain more of the regional benefits of its agglomeration economy.

To some analysts, the Rochester regional story can be neatly summarized as one of institutional "lock-in" (Grabher 1993). In this interpretation, entrepreneurial firms are inhibited from innovative responses to changing global markets because regional institutions have been unable or unwilling to adjust to the possibilities presented by a global economy. Regional institutions and their leadership are locked into patterns, processes, and norms that inhibit innovation. The idea of lock-in as an explanation for lagging regional growth is prominent in AnnaLee Saxenian's (1994) comparison of Silicon Valley and Route 128 in which regional culture plays a prominent role in explaining innovative capacity.

Our analysis of the Rochester region and its agglomeration economy certainly recognizes a "cultural" element but looks at the concrete policies and institutional power that influences how regional actors perceive and respond to market risks and opportunities. The problem with "lock-in" as an interpretation is that, as our analysis demonstrates, the Rochester region and its institutions have adapted to changes in global markets. Rochester is certainly not the same high quality manufacturing region that it was in the 1970s. Recent critiques of the "lock-in" explanation have highlighted similar empirical findings related to the evolution and adaptation of industries in situ and over time (Martin and Sunley 2006, Boschma and Lambooy 2002).

In fact, it could be argued that the kinds of adaptations required of a region such as Rochester, what in the 1970s and 1980s was called industrial restructuring, have been more dramatic than those in Greenfield regions that have sprung up in places like Tucson, Arizona, not coincidentally, the home of another precision optics cluster. Secondly, the Rochester region is home to a sizable group of highly innovative firms serving global markets. What is notable about Rochester is the tension between its ability for institutional and industrial adaptation to shifting global markets and its inability to capitalize on its innovative potential to capture those benefits in the form of higher wages and broader and deeper regional economic growth.

As our case study illuminates, Rochester's trajectory and the changes affecting its workforce are less a function of inability to adapt and change than a question of who is controlling the direction of change and to what ends. Whereas Saxenian sees dominant firms emerging to stifle innovation (a cultural explanation), we see a longer process of dominance by TNCs to direct the trajectory of change in so as to reduce their risks and use the regional resources to their advantage. While it may be an inadvertent result, the result is a lagging regional economy.

As we have shown, the dominant TNCs in the region have, in fact, captured the benefits of Rochester's agglomeration economy, making it difficult for those benefits to be ploughed back into the regional economy to produce higher wages and public investments that could attract and retain a skilled workforce and build a sustainable regional economy built on innovation.

The region we discuss in the next chapter may appear to be light years away from Rochester in its culture and the types of innovative products it produces and exports. We find, however, that because dominant firms in the two regions are established within the same national governance regime they hold strategic dispositions in common. In particular, they have some surprising points of commonality in how they view the regions within which they operate and, in particular, regional labor markets.

5   Runaway production

Media concentration and spatial competition

As Allen Scott (2005) has demonstrated, the Los Angeles region has retained its position as the dominant pre-production, production, and post-production center for the media entertainment industries throughout the twentieth century and into the twenty-first century. That said, production activity outside Los Angeles has been an important factor in both the creative and economic calculus of producers – particularly since the 1970s. However, both the geography of that external production and the factors driving it have changed. With the gradual emergence of a horizontally and vertically integrated media entertainment industry and the changing spatial investment strategies of multi-national firms, the distribution of media entertainment production has developed a new shape.

In the 1980s: the evolution from film to media

"Runaway production" is an old complaint in the media industries. Film crews have routinely left Los Angeles, the historically dominant production center, to shoot in exotic or less expensive locales. However, in the 1970s and especially in the 1980s the pattern of location shooting changed. Film shoots outside the Los Angeles region increased in conjunction with a rise in demand for media entertainment products. This rise in demand was stimulated by the expansion of commercial television in global markets and the emergence of potential new domestic markets such as home video (Prince 2000). The industry rose to the challenge of increased production.

The relationship between production and distribution in the media and film industries has long produced a contested regulatory terrain. During this period of market expansion, the US national government enforced market regulations that fostered competition in distribution and production markets in the media industry via anti-trust decisions and financial syndication rules (Holt 2001). A US Supreme Court decision in 1947 (known as The Paramount Decision) forced the major motion picture studios to divest themselves of their distribution venues – US theaters. Regulated competition curtailed the "Majors'" ability to distribute packages of films through a practice known as "block booking" thus mitigating their ability to manage both their product and its markets.

Because the anti-trust provision did not extend outside the US, however, the major studios retained the ability to book films in Canadian theaters in package deals rather than individually. This effectively limited the development of an indigenous commercial film industry in that country (Winseck 2002). Thus the production side of the US and Canadian film industries evolved differently, in part, because of variation in national regulatory regimes.

Another key regulatory effort to encourage competition and discourage consolidation of production and distribution in the entertainment media industries affected the commercial television networks rather than movie studios. Financial syndication rules, adopted in the 1960s, forced the then three commercial television networks to purchase prime time programming rather than produce it in-house. This led to the emergence of powerful, independent, mid-size production houses producing media for sale to commercial television networks.

In the regulatory environment that shaped the media markets of the 1980s, overall differences in production strategies and product mix among the major film and television product producers decreased. At the same time, the creative differences among individual products accelerated. This produced a convergence in the character of the product coming out of the dominant firms in the industry. Maltby (1998) describes this period as one in which "the post-Paramount attitude of regarding each production as a one-off event had reached a point where none of the Majors any longer possessed a recognizable identity either in its personnel or its product."

During the 1970s and early 1980s, then, independent producers (not principally financed by the major film studios) and mid-size firms, such as Cannon and Lorimar Telepictures, expanded production in response to the increasing demand for differentiated products. For a period, lasting no more than a few years, the bargaining power of the "independents," and their associated production networks, increased vis-à-vis the major studios. This bargaining power was a product of both firm strategies and the regulatory environment. The regulatory nexus created by the Paramount decision and the differentiated products that they were able to produce allowed the "independents" to develop a brief competitive advantage within the industry's production process.

In response, the "Majors" (Columbia Pictures, Warner Brothers, MGM, Twentieth Century Fox, Universal Studios, and Paramount Pictures) used their political influence on regulation and their capacity to re-try the Paramount decision in US courts to re-establish vertical and horizontal integration and re-establish their dominance over media production and distribution (Holt 2001). However, during this brief period of market instability, characterized by increasing demand, new markets, and a variety of product distributors, independent producers thrived in a competitive market for media entertainment products (Christopherson & Storper 1986). It is during this period that the media industry took on indications of a flexibly specialized production model (Piore & Sabel 1984).

Product differentiation spurred technological change. Using television production methods and technology, film-making became more mobile, which enhanced the independence of film crews from the studio and sound stage environment. Film crews were able to take advantage of both the creative possibilities to differentiate products and the lower costs of shooting "on-location."

They also took advantage of an expanding range of incentives provided by US cities and states to lure film crews to shoot their film outside Los Angeles. Incentive packages to lure film crews came to be an expected element and became an additional factor in the production decision calculus. The typical incentive package included inexpensive accommodations for film crews, tax breaks for using local businesses, such as catering, and construction, and easy permitting to use locations including public spaces. States including Texas, North Carolina, Florida, and Illinois vied for two or three films a year, anticipating that there would be pay-offs beyond the short-term jobs created in restaurants, catering, and dry cleaning to serve the film crew. Because the location was part of what distinguished the product, they were valuable as state promotional devices thus influencing popular perceptions of a place. Visibility in a film drew tourists and businesses to places far from Hollywood. The relatively modest incentives were part of community and state image marketing programs, intended to increase the media exposure of the state or city so as to attract tourists and long-term business investment. Attracting film production fell into the same categories of "urban entrepreneurialism" as attracting a sports franchises or building a major league stadium.

The runaway production controversy is often cast in terms of labor costs and the relative expense of Los Angeles or New York City versus other production locations. Labor cost was an issue in the 1980s runaway production controversy but this was in connection with costs at the margins, such as catering and transportation. With the exception of New York City, the specialized skilled labor needed to shoot the film was not readily available in other film-shoot locations. Key members of the production crew were primarily hired in Los Angeles or New York because of their particular skills and connections within a production network. Although studio facilities existed in some of these locations, such as Las Colinas, in Texas, they were fragile operations. The lack of sufficient, consistent production work made it difficult to sustain a skilled workforce in the region. In the rare case that a skilled production worker managed to capture experience through on-location shooting, he or she typically de-camped to Los Angeles in order to build a sustainable career (Christopherson & Storper 1989).

In addition, "pre-production," the development of the product concept and origination of the production crew and financing, soundstage production, and post-production, the editing of the product and finishing for distribution, remained firmly rooted in Los Angeles. These activities were carried out by a skilled and specialized regional labor force, which was highly unionized (Gray & Seeber 1996). The competitive production environment fostered by the regulatory regime encouraged creativity and product differentiation and the regionalization of production in Los Angeles despite high labor and transaction costs.

As a consequence, the skilled media entertainment labor market was little affected by 1980s "runaways." Because the locations where shooting occurred rarely could provide the skilled labor needed to make the film, the cinematographer, script supervisor, or grip went to Vermont or Texas to shoot scenes for a few days and then came home to the San Fernando Valley or Santa Monica to spend his or her paycheck. Similarly, the project-based networks which characterized the hiring process in the industry remained geographically fixed in Los Angeles and New York City even while the production work occurred in a wide range of locations. Runaway production meant job loss only for the less skilled and non-unionized workers who provided localized services: transportation, catering, carpentry, and dry cleaning services to film crews on location (Christopherson & Storper 1985, 1986).

In the 1980s, the most important promoters of "runaway production" as a policy issue requiring regulatory intervention, were the major studios in Los Angeles. These studios had large facilities, high overheads, and a unionized workforce. In other words, they had sunk capital in the region. They pushed for incentive packages in California so as to keep as much production as possible in Los Angeles to fill up their sound stages and to use their equipment and services. New York City was considered the central rival to Los Angeles, the major site for "runaway" production. The labor question entered into this rivalry only tangentially. Labor costs were essentially the same on both coasts because of national collective bargaining agreements (Gray & Seeber, 1996). The choice between New York and Los Angeles was based on other factors. Producers who wanted to work with small production crews and in a collaborative style, for example, gravitated to New York.

The film and media industry remain the quintessential creative economy industry. The value of the product is largely defined by labor inputs such as creativity and talent rather than capital inputs. Thus any debate about cost structures or control in the industry quickly becomes a discussion about the cost of labor. A regulatory regime that encouraged vertical disintegration and a flexibly specialized production industry did not eliminate that conflict between labor and capital or significantly dilute the power of "the Majors" over the distribution of entertainment products. The industry's control of distribution networks in the US and a legal international cartel, The Motion Picture Export Association, maintained US major studios' power over access to the movie consumer market. Thus the line between the industry and its markets remains both blurred and highly dependent on the current regulatory environment.

During the period of the "independents," vertical disintegration and market expansion gradually changed labor market dynamics within the industry in Los Angeles. An increasing number of industry workers in Los Angeles were employed outside union contracts and the "mini-majors," small production and distribution studios, who were expanding at the time, were not unionized and actively fought unionization (Weinstein & Clower 2000). The more inclusive "above-the-line 'talent'" (writers, directors, actors) unions in Los Angeles dealt with the complex consequences of a rapidly increasing membership, including many members with only a peripheral connection to the industry. An expansion of the talent community beyond an elite group led to fractious conflicts among the established industry labor market and the newcomers, and complicated union campaigns to obtain a portion of the profit stream from the products they were producing (Christopherson & Storper 1989; Storper and Christopherson 1985).

In Los Angeles, the media labor unions headquartered in the region began to support studio appeals for incentives by the state of California to stem "runaway production" (for example, providing free police protection to film crews shooting in Los Angeles and clean-up crews, and easy permitting) because of a perceived need to support the regional industry, and because they preferred production close to their homes and families. At the same time, however, they did not see that their interests were directly threatened by shooting on-location. They were protected when working in Los Angeles by "the 30 mile limit" (from La Cienaga and Beverly Boulevards in Los Angeles) within which more restrictive work rules were enforced. Production outside this range entailed additional negotiated benefits for unionized workers and, for the workforce, meant inconvenient travel rather than job loss.

As a result, there was no credible inter-regional competition, including over labor cost. The frictions that existed between East Coast and West Coast branches of guilds or union locals were subdued by national collective agreements and by the dearth of regional competitors with substantial production capacity and a skilled workforce. Policies to stem "runaway" production were, in effect, indirect and rather ineffective strategies to increase demand for use of the fixed real estate (and sunk costs) represented by the major studio facilities in Los Angeles.

In the 1980s, then, media producers used locations outside Los Angeles in limited ways primarily associated with product differentiation. They were dependent, one might even say captured, by their need to use the specialized and collectively organized Los Angeles (and to a secondary extent New York City) media labor market both because of skills and the role of the unions as labor market intermediaries who efficiently assembled reliable project teams. During this period the state played a limited role. State film offices focused on making location shooting easier in order to encourage state tourism. And, while the state of California was more engaged than most in supporting the media entertainment industries because of their centrality to the economy, even in the case of California support was limited to on-location production incentives.

In the early 2000s: the media industry and the role of regulation

Twenty years later, the geography of production in the media industry has changed in response to increases in demand and the expansion of product markets. However, much of what has changed emerges from a new set of firm strategies which are dominated by efforts to renegotiate market risks vis-à-vis both the state and the region. These strategies have resulted in a new geography that illustrates a long-run production strategy which uses inter-regional competition to gain advantage over suppliers. According to evidence developed by the industry, US-based television and film production increased by 36 percent in the decade of the 1990s and has continued to expand into the 2000s (California State Department of Employment 2005; Monitor 1999). Los Angeles has maintained its dominant role in industry production during this period of expansion. Recent work by the Entertainment Economy Institute, for example, demonstrates that production in both feature film and television rose in California in the 1990s, with a 6.6 percent increase in employment (Entertainment Economy Institute & PMR Group 2005). This continued growth took place, however, in a very different production and distribution environment than that which existed in the early 1980s: one in which the small number of media conglomerates have re-acquired ownership of entertainment media distribution channels (cable networks, broadcast networks, and theaters) and are now also able to legally produce their own products for those outlets.

Deregulation and lax enforcement of the market rules that structured competition in the 1970s and 1980s has led to a concentrated media entertainment industry in which a small number of firms control access to multiple distribution markets in film, broadcast and cable television, and DVD sales and rental (Bagdikian 2000; Epstein 2005). As one member of the Creative Coalition, a Los Angeles group of writers, directors, and producers, described the new market, "There are many voices but many fewer ventriloquists."

Six conglomerates – Viacom, Time Warner, NBC Universal (owned by GE), Sony, Fox, and Disney – overwhelmingly dominate the production and distribution of entertainment products in the US in the early 2000s. Together they control 98 percent of the programs that carry commercial advertising during prime time television (including commercial network and cable programming) and 96 percent of total US film rentals. They control 75 percent of commercial television in none-prime-time slots, 80 percent of subscribers to Pay TV, and 65 percent of advertising revenues in commercial radio (Epstein 2005). Although one might expect some competition within this oligopoly, inter-firm competition is minimal. The firms cooperate to influence policy (such as that compelling manufacturers of DVD players to implant a circuit that would prevent playing movies from Europe in the United States) through their trade association, The Motion Picture Association of America (Epstein 2005). They have also formed alliances within the group of six to reduce their risks, by coordinating the release of products, and to ensure that they all have maximum access to global markets (Epstein 2005). The control of end markets by a small number of media conglomerates and the combination of lower risks in distribution markets and need for production volume has translated into a production process that is, if not vertically integrated, "virtually integrated" (Christopherson 1996, 2002)

Concentration has altered all aspects of the production and distribution of entertainment products, including the cost structure of production, what is produced, and the production process itself. At the heart of this restructuring is the ability of "virtually integrated" transnational media firms to use (or in industry jargon, to "repurpose") products across multiple distribution outlets. This capacity multiplies the revenues that can be obtained from any one product – film, television series, or documentary – as it is repackaged, redistributed, and resold as a "new" end product. It also reduces transaction costs and direct costs associated with product acquisition for downstream distribution venues, such as DVD.

The integration of production and distribution has spawned new strategies to squeeze more profits out of media products. One example of these new strategies is cross-market advertising. Films produced by Disney are advertised on its ABC television network and promoted through its news and information programming on its various programs and channels to create a media buzz and hype. Theaters owned by the conglomerate advertise its other media products. Television and film products are bundled by the conglomerate for sale to ancillary markets such as in-flight entertainment packages for airlines. As media markets continue to expand and differentiate through music, the internet, print media, radio, films, video, and television, these bundling and cross-promotional strategies become increasingly ubiquitous.

Media conglomerates are now able to avoid the expensive, transactions-intensive process of buying products from independent producers by producing products within their wholly owned subsidiaries. Unlike many other traditional industries in which firms have embarked on vertical disintegration strategies to develop networks of suppliers they control through cost-cutting and competition, the film industry has consolidated its supply chain to strategically control its production process and its expanding markets. This strategy not only limits the risks and uncertainty of a volatile consumer market governed by short cycles of popularity and taste, but provides a mechanism for controlling the cost of production inputs throughout the production process. Further, consolidation and virtual integration establish high barriers to entry for new firms (both international and domestic) attempting to compete in a lucrative and expanding market.

The 1995–96 US television season was the last in which broadcast networks were restricted through regulations from producing their own programming. At that time the networks had at least partial ownership in under 20 percent of their new shows. In 2002, the television networks, and the media conglomerates of which they are a part, had increased that percentage to 77.5 percent (Manly 2005).

This restructuring of the media entertainment industry has, in turn, affected the bargaining relationship between capital and labor. It has also affected the location of production activities. Unlike the previous model, the decisions made now regarding where to produce a film or a television program are more likely to be based on cost criteria rather than from a product differentiation strategy.

The feature film remains at the top of the entertainment media product pyramid, with respect to both costs of production and potential profits. The stakes for feature films, in terms of risk and reward, remain high. Feature films are released in theaters before being distributed through numerous other distribution venues and thus live a long life of "re-purposing." Thus, feature films remain an illustrative example of how production and distribution has changed and the impact of industry concentration.[1]

The cost of producing and distributing a feature film has increased dramatically in the last twenty years along with concentration in the industry (Jones 2002; Motion Picture Association of America 2005). Two potential reasons can be advanced for this increase. First, the media conglomerates depend on movie stars as a risk reduction and value-added strategy to increase potential profits across the multiple distribution markets they own. Second, costs have increased because of expanded marketing on the various platforms, including network and cable television, in which the product will be distributed. Because of this cost structure, which requires a significant amount of capital up-front, production and distribution of a film requires multiple financial partners. The media conglomerates (who control the distribution gateways) frequently assume only a minor position in the investment. Producers must engage in complex co-financing deals, looking for finance capital willing to take a risk, wherever they can find it. This causes them to look for regions which will provide them with production cost breaks or direct subsidies which mitigates not simply their costs, but their costs at the beginning of a project.

The efforts to reduce increasing expenses, while largely a consequence of the "star strategy" and the mounting cost of extensive product marketing campaigns, has instead focused on "below-the-line" or skilled craft labor costs. As in many outsourcing stories where skilled labor becomes the focus of firm cost-cutting strategies, the craft labor engaged in film production is perceived as less important in adding value to the product and acquiring necessary financing. Financial partners want to know who the star will be or they want to see the marketing plan before they invest in a project. They are generally less interested in the skills and experience of the project team assembled to design, shoot, and edit the film although many variable costs indeed depend on the competence, quality, and creativity of project team.

As media concentration has proceeded, the increased bargaining power of conglomerate distributors vis-à-vis producers and the media industry labor market has produced dramatic changes in entertainment media working conditions and labor relations. Most of the pressure has been felt among the below-the-line labor market in the form of increasingly unpredictable employment and mandatory overtime. The variability of the production process is thus transferred to the labor market.

The changes wrought by industry concentration affect even the most creative segments of the industry, however. And the porous firm boundaries, which have been strategically renegotiated through consolidation and concentration of production and distribution, make the old rules governing compensation for the creative work product seem antiquated. According to one veteran film-maker:

> in cable, residuals for writers, actors, and directors are a percent of the producer's gross. But if that producer is a network who self-deals the rights to their cable company . . . there is no compensation for that. Suddenly you discover that the eleven or twelve per cent gross residual among the three guilds that has been fought over for so many decades is virtually meaningless, as rights are simply self-dealt among related entities,
>
> (Hill 2004)

In the media industry, as with so many industries, the psychological contract has shifted gradually, and new rules have yet to emerge. Even in an industry based in creativity, debates over appropriate compensation and ownership for "intellectual property" and "talent" remain unresolved and the previously negotiated arrangements are eroding.

Virtual concentration and the rise of inter-regional competition

As the bargaining position of media conglomerates with labor has changed, however, so has its bargaining position with the regions. These regions are the locations in which shooting and, in some cases, sound stage work, take place. As Galbraith (2004) describes in examining the production conditions for all innovative industries, skilled labor is critical to the production of entertainment media. So, for the media entertainment industries, labor relations and location strategies have become intertwined.

Since the late 1980s, production of some types of media entertainment products has taken place outside the United States, particularly in Canada but also in the United Kingdom, Australia, and much more occasionally, in Eastern Europe. These changes in the spatial location of production are often portrayed as "runaway production." The competitor to Los Angeles is no longer New York, as it was in the 1980s, however, but international production centers with sound stages, equipment companies, and, most importantly, cheaper craft or below-the-line workers. By contrast with the 1980s, when runaway production referred to shooting on location, soundstage production and post-production editing and special effects production, as well location shooting are occurring in these production "satellites," most particularly in Vancouver, Canada (Coe 2001). This "runaway production" is thus of a different character than that at the center of policy debates in the 1980s. However, the media industry is also a different industry, engaged in a very different market than it was in the 1980s.

There have been other shifts in the "runaway production discourse" that reflect changes in the power relations among the media conglomerates, the producers who provide them with products, and the media labor market. In this round of "runaway production" policy debates, the major studios, although headquartered in Los Angeles, are part of production and distribution transnational media conglomerates. As a result, they are now on the other side of the runaway production debate. They now support the "right" of producers to produce where they choose, not just outside of Los Angeles and New York City, but also outside the United States.

And again, by contrast with the controversy of the 1980s, labor unions representing media workers in the US are deeply engaged in the contemporary runaway production policy debates. Since the late 1990s, the Canadian government has been accused of unfair competition by US-based media unions, some media service firms, and small non-conglomerate studios, such as Raleigh in Los Angeles (Film and Television Action Committee 2004; US Department of Commerce 2001). At the heart of this charge is Canadian national and provincial provision of labor-based subsidies to US entertainment media conglomerates to encourage them to not only shoot on-location but also use sound stages to produce in Canada. This subsidy strategy attracts more that on location shooting but also the production work that is at the heart of the specialized regional labor markets in Los Angeles and New York City.

For their part, the media conglomerates support the Canadian government's rationale for subsidies to what they refer to as service production, based in the "cultural exception," a provision of international trade law that allows governments to provide subsidies to encourage the production of media products that sustain cultural identity. As in the case of the Paramount decision's critical role in determining the allocation of risks in the production process and the corresponding geography of production, the Canadian privileging of the film and media industry as "cultural production" produces another geography, based a new set of cost structured and negotiated relationships.

The Motion Picture Association of America, the trade association which represents the media conglomerates, fully supports the Canadian subsidy structure:

> Trade action against Canada "threaten to further sour US–Canada relations already strained by tariffs on Canadian lumber, as well as hurt efforts to dismantle barriers abroad to US movies . . . It's a "dagger-to-the-heart challenge to very sensitive cultural subsidies," said Richardson, whose group represents Hollywood Studios including the Walt Disney Co., Sony Corp.'s Sony Pictures and AOL Time Warner's Warner Brothers.
>
> (Pethel 2002)

Again, by contrast with the runaway production discourse in the 1980s, media labor unions in Los Angeles have been the loudest to decry a new wave of "runaway production." However, evidence indicates that the industry continues to grow in the Los Angeles region even as the level of Canadian production increases (California State Department of Employment 2005). As conglomerates have strengthened their control over the industry and as the international market for entertainment products has grown, Los Angles has captured a disproportionate share of the expanded production activity (see Figure 5.1). The Los Angeles advantage, however, is in part New York City's loss. Since the late 1980s, there has been a steady movement of firms that produce commercials from New York City to Los Angeles. This movement responds to the increasing concentration of advertising dollars controlled by the virtually integrated media conglomerates headquartered in Los Angeles.

It is in US production centers outside Los Angeles that the impact of big media's ability to turn to a cost based, rather than content based, location strategy is most apparent. These locations formerly benefited from location shooting for the purposes of product differentiation and have been the big losers in an environment in which cost trumps creative imperatives in making location decisions. Independent film-makers continue to make films with the look and feel of particular places but these films are a very small segment of total film production and, if successful, serve the prestige needs of the media conglomerates to boost their credibility as producing art rather than their profit goals associated with producing entertainment product.

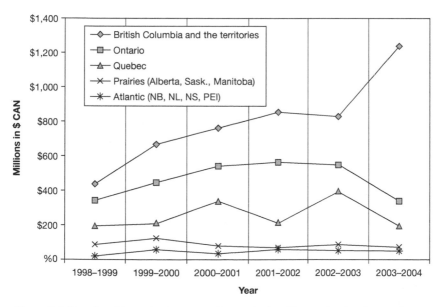

Figure 5.1 Canadian film and television industry: foreign film production by province

By the end of the 1990s, states such as Texas, North Carolina, Florida, and Illinois, which had developed a small media industry base by providing technical assistance to film-makers shooting on-location, some studio facilities, and modest incentives, suffered dramatic declines in production shooting (Jones 2002; Lee 2002). The interpretation of this trend did not focus on firm strategies or on changes in what was produced and how it was produced, but rather on the ability of places and regions to supply the needs of a global industry. Most analyses of the increasing cost-based character of production location decisions was attributed to the rise of *international* competition based on the ability of regions in the US and abroad to provide both the physical and human infrastructure that supports the industry (Weinstein & Clower 2000).

As a consequence, states and cities in the US were pressured by regional economic development coalitions to find ways to "level the playing field," to move beyond the public service incentives they offered in the 1980s to provide direct subsidies, studio facilities, a skilled labor force, and production financing to media producers in order to keep them in the United States (Entertainment Industry Development Corporation 2001; Jones 2002; Monitor 1999; US Department of Commerce 2001).

In the early 2000s, the most generous subsidies were provided by the ill-fated state of Louisiana, one of the poorest in the US The Louisiana subsidies offset up to 17 percent of a film or television production budget and there were no caps on the amount of subsidies the state provided to media conglomerates. In her state of the state address in January 2005, Louisiana's governor, Kathleen Babineaux Blanco said that Louisiana's $73 million dollar growth in tax revenues in 2004 would be unavailable for teacher raises because it would, instead, fund the (currently) $70 million film tax credit program (Webster 2005). In Louisiana's program, investment tax credits are sold at a reduced rate to individuals and businesses, which use them to reduce their own tax liability. The cash goes directly to a production company, such as Twentieth Century Fox (owned by the conglomerate Fox News), a frequent user of the program (Miller *et al.* 2001). The tax credits are typically bought and used by professional business services, including the law firms representing the media conglomerates operating in Louisiana. A cottage industry has grown up around tax credit sales. One of the major firms in this new tax credit sales industry has established a non-profit organization to "unite the players in the state's entertainment industry (Miller *et al.* 2001) to lobby the state to maintain and increase the subsidies supporting the location of productions in Louisiana. Their goals are to obtain public capital to support the construction of a soundstage (Louisiana currently has one) and to monitor what other states are providing so that Louisiana can continue to match and exceed these bids (ibid.). The state is also being encouraged to invest heavily in training programs to provide Louisiana with a skilled media entertainment workforce, which it currently lacks.

The Louisiana story is but one example of how state and national policy seeks to shape the geography of the film industry through regulation and incentive structures. Media conglomerates, well aware of the competing efforts by states, regions, and nations to attract production, strategically incorporate government incentives into their cost structures and locational considerations.

Some of the key actors lobbying for direct subsidies to transient production companies are similar to those who spurred location shooting incentives among states in the 1980s. Real estate interests, for example, play a critical political role in a growth coalition that stands to benefit from an increase in location shooting. At the same time, both the political arguments for supporting location shooting and the character of the support requested have changed significantly. In the 1980s, states invested modestly in film commissions and incentives to make it easier to shoot films (through eased permitting and location assistance) hoping to attract media attention. Incentives increased in the 1990s to include sales tax exemptions and waivers on hotel occupancy taxes but since these incentives constituted a minuscule portion of the shooting budget, they had little impact on location decisions.

In the new economic development environment regional growth coalitions (and their industry allies) argue that in a global marketplace in which all regions are competing for industries with good jobs, public capital plays a critical role in creating the conditions within which occasional film and television projects will be transformed into a stable and lucrative regional industry.

The scenario follows the well-worn path of stage theory. First, industry projects must be lured to a region by significant direct financial subsidies, demonstrating its good business climate. Second, the region is charged with developing and sustaining the infrastructure, including capital facilities and a skilled labor force, to sustain industry "presence." In the case of media industry shooting in Louisiana, the state is also charged with organizing a skilled labor pool from other states that can come into Louisiana to work on productions receiving subsidies – encouraging a "brain drain" from other industry centers. The rationale provided for the munificent subsidies is that they will lead to the development of a Louisiana-based creative economy industry. Canada is used as an exemplar of this process and its success is attributed to the initial stimulus of subsidies. An examination of how the "model" Canadian media industry has developed over time, however, suggests that development policies oriented toward providing subsidies, skilled labor services and facilities to transnational corporations have uncertain and complex consequences even for high-skilled industries.

Servicing the transnational media industry: the Canadian case

According to the US industry story, the movement of significant entertainment media production to Canada, particularly for network and cable television, is attributable to the introduction of Canadian production tax subsidies in the 1990s. The line of argument is that production costs rose in the 1990s and that media product producers looked for ways to reduce below-the-line (production crew and craft worker) costs. They identified Canada as an alternative because Canadian labor-based production subsidies reduced below-the-line costs. In 2003, for example, Canadian national subsidies to foreign producers, in the form of tax credits, were increased from 11 percent to 16 percent of labor costs (Canadian Film and Television Production Association 2003). Canadian national subsidies are enhanced by additional tax credits provided by Canadian provinces competing against each other to attract foreign production. The total tax credits available to US producers on payroll costs are as high as 44 percent (Blackwell 2003).

As a consequence, so the argument goes, production of television pilots, some series, and made-for-television movies began to take place in Canada (Jones 2002; Monitor 1999). Subsidies are given credit for attracting production away from US regions, including Los Angeles, and building a Canadian production industry that can compete with the US industry. To compete with Canada, then, the US must subsidize entertainment media production at the state and federal level to bring US labor costs in line with Canadian labor costs. States wanting to build an entertainment media industry are provided with a trajectory of Canadian success, which begins with inter-regional competition via subsidies to attract transnational firm production with the lure of low labor costs and ends with a successful free-standing capacity to compete in the global entertainment media industry (Monitor 1999). A prominent New York economic development official tied the development of a successful and prototypical Canadian industry to the subsidies the Canadian government provided in the early 1990s. "Before those subsidies, producers were not interested in Canada." According to this supply-side explanation, the subsidies worked and "in a few years Canada had a developed competitive media industry."

This argument is compelling to regional growth coalitions because the costs and risks of industry development are assumed by the state rather than the private sector. It is appealing to unions because advocating for state support for jobs in the region is an easy way to promote guild and union membership among the increasingly large portion of the entertainment media production workforce that does not belong to unions and may, in fact, be hostile to them because of their gatekeeper function.

Actual facts concerning production trends across international media production regions, however, raise serious questions about the accuracy of this model of industry development. They suggest that it is more powerful as a narrative to promote inter-regional competition than a prescription for the development of regional industrial capacity.

First, there is the question of the relationship between growth of production in Canada and employment trends in the US The so-called rise of the Canadian media production industry does not coincide with decreasing employment in Los Angeles and New York, the major US production centers. Employment grew in California and New York in the 1990s despite increasing Canadian labor-based production subsidies (California State Department of Employment 2005). And Canadian production has stagnated in the early 2000s despite Canadian national and provincial increases in subsidies to producers (ibid.).

Employment patterns in the two major US centers appear to be more effectively explained by the ups and downs of the business cycle and to fluctuations in the value of the Canadian dollar (Entertainment Economy Institute & PMR Group 2005) than by Canadian tax-based subsidy packages.

Second, producers could not have moved production to Canada in the 1990s were it not for the presence of a developed industrial base, including a skilled labor market and production facilities especially in what became the satellite center of Vancouver. With the shift in the factors driving production location decisions from product differentiation to production cost, producers on contract to the major media conglomerates were required by their employers to move production in Canada for cost reasons. While some "beauty shots" might be made in New York or Chicago where the film was actually set, the expensive sound stage work was transferred to lower cost facilities in Toronto and Vancouver. In the case of Vancouver, sound stages were leased by the media conglomerates so that they might be used continuously by US-based producers.

Third, the presence of an industrial base that could be used by the media conglomerates was not stimulated by international outsourcing by the media conglomerates but the result of long-term investments by the Canadian state.

There are, in fact, a wide range of ways in which Canada has supported the development of a media production industry across the country. Some of them are an outgrowth of the power of Canadian labor in crafting a national welfare state that provides Canadian citizens with universal access to health services and a generous (compared with the US) educational system that produces a skilled workforce. US producers locating production in Canada are subsidized indirectly by investments made by Canadian taxpayers to build their national and regional production capacity and provide for the social security of Canadian citizens.

Other "subsidies" reflect the importance Canadians place on preserving a portion of programming that demonstrates a Canadian point of view and supporting production with Canadian content. Over time, the inability of Canadian independent film producers to make a dent in oligopolized film distribution in Canada (Winseck 2002) moved them to turn their attention to television, which was still state-supported to promote an arena of Canadian content. In 1984, the Canadian Film Development Corporation became Telefilm Canada and since 1988 has invested more than $60 million annually in television programming. Government expenditures to support film and television production in Canada are significant, representing 10 percent of the total yearly budget for cultural activities (Montpool 1998). This support focuses almost exclusively on television because continued Canadian state television distribution assures a distributor, a prerequisite to receiving a production subsidy. In 2003, productions made with Canadian content received a 60 percent domestic tax subsidy on qualifying labor expenditures (Canadian Film and Television Production Association 2005). The Canadian Television Fund has been important in encouraging regional production in the Canadian provinces and thus developing an industrial base throughout the country.

The Canadian media production labor market is particularly trained to produce programming for television. In addition, taxpayer-supported investment in Canadian television production has fostered the development of a skilled media labor market across the Canadian provinces, not just in the major media centers of Toronto and Montreal. The Canadian government supports training and mentorship programs administered by the CFTPA, which also works with Canadian producer-distributors to train Canadian media workers (Canadian Film and Television Production Association 2005).

Finally, an interpretation of the reasons for international outsourcing in the media entertainment industries is inadequate without considering the differences in media industry development across the Canadian provinces, particularly Ontario and British Columbia. Regions in Canada play distinctly different roles in the emerging international division of media labor. Toronto has historically played the central role in national media production, both in television and independent film. Politically, this has translated into constituencies both for and against subsidies to foreign producers. Those who oppose the support of foreign (US) service production argue that it actually undermines regional capacity to develop a competitive position in the global media market because it uses resources that could go to Canadian production to serve the interests of foreign producers. One spokesperson for this view is Telefilm Canada executive director Wayne Clarkson: "Building an industry based on foreign production is like building a house on quicksand. A strong indigenous production industry should remain our prime directive."

While Ontario and Quebec, and to a lesser extent the remaining Canadian provinces, are competitive sites for on-location shooting during Hollywood expansion cycles, Vancouver and British Columbia can be described more accurately as a Hollywood low cost production satellite (Coe 2001; Elmer & Gasher 2005). The use of this satellite would have been inefficient before media concentration because the major media conglomerates did not control the distribution outlets that now make it less risky to turn out products that can be "repurposed" again and again across multiple distribution channels. Virtual integration made it possible and economically efficient to make long term investments in sound stages in Vancouver so as to churn out batches of lower cost productions. This works for Vancouver as long as it remains a low cost site but there has been only marginal progress in transforming the City into a free-standing globally competitive media production center.

For, while Canada has made substantial investments to attract US producers and Hollywood has responded by investing in facilities in British Columbia, there is also evidence that the "bargain" is fragile. In 2003, as the Canadian dollar's value rose against the US dollar production began to decline. (The

Canadian dollar hit a 12-year high of 85 American cents in 2004, almost 30 percent higher in value than in January 2003.) Although the decline in 2003–4 was small, the changing conditions manifested in much more stark terms how the transnational media firms were using Canadian facilities and labor.

The numbers show perilously dramatic swings in work among the regions (see Figure 5.1). British Columbia saw an almost 50 percent increase in foreign location production while Quebec dropped by 50 percent and Ontario dropped almost 40 percent. Preliminary numbers for 2004–05 show across the board decreases. More recent evidence suggests dramatic declines in British Columbia, particularly in the number of projects that are locating in Canada. British Columbia now fears losing its skilled labor to Los Angeles where the core jobs still remain.

In sum, demand trumps supply as an explanation for what has happened in the Canadian media entertainment industry since the early 1990s. It is the changing organization of production in the media entertainment industries that allowed them to effectively utilize the investments that Canadian citizens had made in developing regional production bases over a period of fifty years. The recent tax-based subsidies although they contribute to the attraction to Canadian centers are a consequence of vicious inter-regional competition within Canada. This inter-regional competition and its gradual extension to US states and regions increases the profits of transnational firms rather than building competitive regional industries.

Interpreting the story of inter-regional competition in the media industries

As we argue throughout this book, transnational firms using skilled labor that produces creative or innovative products face a particular set of problems. The first of these problems is the reproduction of a sufficiently large and skilled labor market, outside of the historically sustaining framework of the firm. The second problem is control of the wage bargaining power of skilled workers in an unpredictable labor market. The third problem concerns firm needs to obtain flexible production conditions, including in hours of work and in working conditions – what are generally called "work rules."

In looking at how these "problems" were solved in the media entertainment industries in the 1980s and how they are solved in the early 2000s, we can see the impact of changes in both inter-regional, and inter- and intra-firm competition. Simply put, media concentration of power has paralleled the rise of inter-regional competition.

We do not argue a causal relationship between these two globalization processes but rather that they have intersected with the balance of power shifting from labor and regional industrial production complexes to the transnational firms. In the media industry, entertainment media conglomerate control of distribution gateways has given them enhanced control over what is produced and how it is produced. TNC control over the most predictable types of entertainment media production, such as episodic (series) television has increased TNC bargaining power vis-à-vis labor and the regions. In an inherently risky business, conglomeration (the manifestation of concentrated power) has enabled TNCs to shift risk and the cost of sustaining a project-oriented creative workforce to labor and the regions. At the regional scale, the rationale given for taking on these burdens is that they will build industry capacity and regional comparative advantage. Evidence from the media industry suggests, however, that inter-regional competition over TNC production undermines the ability of regional complexes to develop or sustain distinctive industrial specialization.

Some of the reasons behind the change in the balance of power between transnational firms and the regions are well documented. They can be traced to an altered international trade environment in which TNCs move more easily across national borders, differentiating among various kinds of economies and the comparative advantages they provide (Gereffi *et al.* 2005; Kogut 1984, 1985).

At the same time, the devolution of national responsibility for social welfare and economic development to the regional scale has heightened tendencies toward "disorganized" or competitive capitalism, including at the regional scale (Brenner 2005; Lash & Urry 1987; Offe 1984). The emerging international division of labor is most certainly affected by these changes in production alternatives and in the pressure on regions to self-finance infrastructural development and social protections. The changing power balance between TNCs and regions has been shaped, however, by two other processes intimately connected with contemporary globalization: 1) the promulgation of an ideology of endogenous regional growth (Lovering 2001; Martin & Sunley 1997) and 2) changes in national regulatory regimes that have concentrated economic power in TNCs under the banner of creating globally competitive national champions.

Conclusion

The story of how the locational strategies of media production and distribution firms have changed between the 1980s and early 2000s contains some valuable insights into the forces and processes shaping the particular form of global economy that is emerging.

A literature on the new international division of labor stretching back to the 1970s has demonstrated that the ability of transnational firms to exploit comparative advantage in wage rates and labor power has always required collaboration among transnational firms, regional property developers, and the national and local state. Without modern infrastructure, transnational firms find it difficult and inefficient to take advantage of comparative differences in wage rates. The first enterprise industrial zones, distinctive enclaves in which the telecommunications and transportation infrastructure was made to order for transnational firms or their supplier, were critical to the construction of a new international division of labor in low-skilled manufacturing. International competition among these manufacturing zones was critical to the ability of transnational firms to drive down wage rates while maintaining the logistical infrastructure to enable just-in-time production and distribution.

The outsourcing of skilled work, especially that requiring innovation and creativity requires even more intensive use of state resources, not only in infrastructure but in the provision of a skilled workforce. Behind these public investments are those who will particularly benefit from enclave development – property developers and those firms servicing the transnational enterprise. In his paean to the end of geography, Friedman (2006) describes a luxurious "campus" in India, with putting greens and swimming pools for transnational corporate executives and skilled workers who provide services for low wages to transnational firms. This comparative advantage comes at a significant cost, however. Outside the "campus" is a world of pockmarked roads, horse-drawn carts, motorized rickshaws, and barefoot drivers (ibid.) which Friedman notes but doesn't recognize as integral to the vision of the global world he is constructing. The campus enclaves represent a line of investment strategies stretching back to the early days of global production, perhaps quantitatively but not qualitatively different from what has preceded them.

The existence of these low wage skilled enclaves suggests, however, that low cost labor, while it may be necessary to transnational investment, is not sufficient. Comparative advantage derives not just from an aggregation of low wage workers but from the existence of the conditions that make it possible to exploit those low wages to extract increasing returns. This requires public investment in infrastructure and national and regional cooperation to create agreeable labor conditions. If comparative advantage is not rooted solely in the presence of individual low wage-workers but depends on public investments and regulatory environments, we need to focus attention on the implications of those regulatory environments and investments. It is in this context that we begin to see the two critical processes shaping the global economy. The first is the fragmentation of work, including the separation of more routine, "bread and butter," production from higher risk, less predictable production. This process not only separates classes of workers from one another but increases the vulnerability of the workforce at both the more routine and highly creative ends of the production spectrum. A second key process is that of intensified and fostered inter-regional competition. As promoted by transnational firms and their trade associations, this competition encourages replication of certain types of skilled labor and facilities across regions rather than specialization and niche production. It thus potentially undermines regional competitiveness based in specialized skills by neglecting investment in the specialized assets of the region.

Section III

Learning regions and innovation policies

6 The paradox of innovation

Why regional innovation systems produce so little innovation (and so few jobs)

The concept of "innovation systems" emerged in an important branch of the strategic management literature that addresses industrial transformation in advanced economies. These "systems" are composed of networks of firms and the institutional infrastructure that supports them. Two premises underpin the idea that innovation systems can serve as the basis for an economic development strategy not only in advanced economies but also in emerging economies that hope to "leapfrog" into knowledge-based industries.

The first emanates from a broad understanding that "the economy has shifted from a labour and capital-based economy to a knowledge-based one, where knowledge is the most important resource and learning the most important process" (Boekema *et al.* 2000: 4 cited in Sokol 2003: 56). Technological change and particularly changes in access to and the cost of information technology are assumed to lie behind this post-industrial revolution. Because technical advances are critical in the knowledge economy, continuous innovation is presented as necessary to firm competitiveness.

A second premise is that interaction (and competition) among a network of inter-related firms will foster innovation and accelerate the pace at which it occurs (Porter 1998). By contrast with the learning region approach, which we will discuss in the next chapter, the region is a secondary player in innovation systems, which are focused squarely on the network of firms. Regions enter into the concept indirectly via a connection with what Allen Scott described as industrial spaces. In these spaces, agglomeration economies are achieved via vertically disintegrated production systems that have recomposed into an inter-related network of firms tied together by their use of shared infrastructural resources and a labor market (Scott 1988).

In this chapter we expand on our analyses of firm strategies and labor markets in the first chapters of this book to examine how regional innovation systems are paradoxical. They offer the possibility of economic growth and job creation but, at the same time, that promise is rarely fulfilled. Even within reasonably successful systems, such as those we have described in Rochester in the photonics industry, and in Los Angeles in the entertainment media, real (that is, market altering) innovation is rare. More typically innovation is confined to a narrow range of product development, for example, in distribution technologies, or takes the form of process innovation. In addition, the construction of knowledge (and the processes through which it is diffused), serve the purposes of some firms more than others. This paradox raises questions about the innovations system framework – its piecemeal integration of non-economic concepts, such as trust and cooperation, to explain the functioning of the network; the absence of a critical stance on what constitutes innovation, and its failure to consider the role of power in structuring how firms interact in networks.

Finally, we consider the line of reasoning that connects innovation systems to successful regional economies particularly through the development of livelihood-sustaining "good" jobs. Although analysts of innovation systems rarely concern themselves with this relationship, the association is made in the public policy literature to legitimate government action to support the innovative network agenda. For example, the website for the innovation systems "spin-off," the Council on Competitiveness asserts:

> In our global economy, place matters more than ever. Even as technology, capital, and knowledge diffuse internationally, the levers of national prosperity are, in fact, becoming more localized. As talented people and new ideas become the most critical drivers of economic growth, regional economic conditions have assumed greater importance. Regions that can attract talented residents and support the development of highly innovative firms will support great prosperity. Regions that rely on low-cost labor and basic extraction of natural resources will not. While the US has many successful regions, America is also home to many areas that do not offer the environment necessary to support productive firms – and the higher salaries those firms offer. We are becoming a land of innovation haves and have-nots.
>
> (Council on Competitiveness 2007)

Three assumptions underlie this statement: 1) Innovation networks are exclusively regional and public investment in the network will result in economic benefits to the region; 2) Firms using a high proportion of skilled workers produce new products and services (rather than simply maintaining products already commercialized or replicating established product formulas); and 3) Innovative firms produce new jobs in the region, particularly high skilled, high wage jobs.

In Chapter 5, we focused on the media entertainment industry, showing that transnational corporations, in our case the entertainment industry conglomerates, can tap pools of skilled workers across regions. The presence of network connections across regions raises questions about what portion of public investment in TNC-dominated networks actually accrues to the region that provides that investment and what portion is siphoned off to boost the bottom-line of TNCs.

In this chapter, we look at the assumed relationship between innovation systems and regional economic development from an intra-regional perspective. We examine how the use of a skilled workforce isn't necessarily about the kind of product innovation that produces new jobs. Instead the increased use of skilled workers may reflect a process innovation strategy that reduces the numbers of jobs created. Or, it may reflect the use of pools of "brain power" to service already commercialized complex systems or, as in the entertainment media, to replicate already established formulas for commercial products. In all these cases, skilled labor does not equate with product innovation or with significant job creation.

To begin this discussion we look at how the region has become part of the way we think about innovation systems made up of firm networks.

Regional innovations systems and the regional project

The connection between vertically disintegrated production systems and spatially co-located networks of firms has been a project of economic geographers rather than of either management theorists or economists, for whom the regional dimension is an afterthought.

The primary focus of the *new economic geography* has been the question of generating and promoting increasing returns to scale through agglomeration economies. However, the theory of increasing returns to scale is problematic in classical economics, in which most models assume a constant returns world. "The basic problem with doing theoretical work in economic geography has always been that any sensible story about regional and urban development hinges crucially on the role of increasing returns" (Fujita *et al.* 1999). This tension is at the heart of the new economic geography, which, like regional science before it, seeks to foster and establish a dialogue between economics and geography (Barnes 2004; Glasmeier 2004).

Fortunately, for economic geographers and policy-makers, increasing returns to scale are well documented (Isard 1956; Marshall 1920). The empirical evidence leaves little room for doubt that agglomeration economies play a crucial role in determining costs, productivity, and locational choice. Recent discussions of spillover effects have been built on this foundation. The influence of these processes is significant enough to affect decision-making behavior on the part of firms and industries and the distribution of economic activities. But this theoretical uncertainty surrounding increasing returns underscores a challenge in economic geography that does not afflict classical economics: moving from evidence to theory rather than theory to evidence. In economic geography the real world matters.

So, in economic geography, the discussions of increasing returns to scale, endogenous growth, and agglomeration have been influenced by empirically informed theories about the scale of regulation and the role of the state (Morgan 2004). Much of what has emerged from this dialogue has fixed on the region, particularly the city-region, as the primary scale of analysis in economic geography. The intersection between regional innovation systems and economic geography lies in a mutual interest in the emergence of "knowledge-based" industries and in the factors which attract and sustain them. In both literatures, innovation is, at least indirectly, linked to human capital through a concern for entrepreneurship, research, and innovation (Dunford 2003; Feldman *et al.* 2001; Glasmeier 2002; Hanson 2003).

The attempt to conflate the firm-oriented innovation networks with the region has been severely critiqued (Martin & Sunley 2001). This critique points to the selective and piecemeal way in which non-economic concepts have been introduced to explain the presence of innovative firm networks and how they function. Critics also point to what has been missing. Regional innovation systems are disassociated from broader macro-economic and regulatory policies that steer investment in spatial territories and the regional political economy is reduced to "milieu" or "entrepreneurial ethos." Finally, and central to our analysis in this chapter, a conflation is drawn between successful regional innovation systems and regional economic development.

For example, proponents of innovative firm networks point to knowledge diffusion as a prominent exception to the economic treatment of interdependencies among firms. The recognition that the interactions among networked firms move beyond economic transactions to a sphere of untraded interdependencies has connected the concept of regional innovation systems to broader more explicitly institutional approaches to addressing the challenges of the so-called knowledge economy (Storper 1999). This extension expresses the political-economic institutional context within which the network functions, but only in a limited way. So, knowledge diffusion is analyzed as it occurs among firms in the network rather than as a more broadly constructed and institutionally based collective resource, as it appears in some variants of the learning regions concept. And trust and cooperation are considered as characteristic of individual firms interacting with one another apart from the politically constructed incentives that enable or discourage cooperation among economic actors, including networked firms (Christopherson 1999).

In the policy arena, the regional innovations system approach is particularly consonant with U.S style, or more broadly, Anglo-American, market governance rules because it is driven ultimately by the competitive strategies of firms in global markets. By comparison with the learning region approach discussed in our next chapter, regional innovation strategies are industry driven rather than industry focused. They conceptually disassociate firm competitiveness from regional economic sustainability except to argue that regions that do not provide the conditions to make firms competitive in global markets are destined for failure.

In the next section we critically examine two key premises of the innovation systems framework that underlie arguments for economic development strategies focused on industry-based firm networks. The first premise is that innovation systems foster diffusion of innovation throughout a network of co-located firms. The second is that network firms share in the spill-over effects produced by co-location, particularly labor force skills.

The role of power and the limits to knowledge

Regional innovation systems lead to circumscribed forms of knowledge diffusion and limited types of innovation. To explain why this is so we need to understand that the knowledge economy is only secondarily about knowledge. "Knowledge" is created, and its creation and application can be directed to serve the interests of powerful players. As we laid out in Chapter 2, firms whose primary objective is short-term profit and long-term sustainable competitive advantage will have a strong interest in constructing knowledge creation and diffusion systems in their own interests. This does not mean that they are always prescient or successful but that they are powerful influences on what knowledge is created and where it goes.

As our case studies of photonics and media entertainment suggest, firms with more market power, particularly TNCS, have more power to construct what constitutes knowledge and, by extension, the type of regional innovation system that will support their goals in national and global markets. Ultimately their objectives and strategies may conflict with projects to develop a sustainable economy. For example, in our case study of Rochester, TNCs undertake strategies to provide for their own labor force needs – for a flexible, low cost workforce – and undercut small firm needs for a long-term, multi-skilled workforce. Or, in the media entertainment industry, the dominant players attempt to ensure their long-term sustainable competitive advantage by preventing independent film makers from getting their "innovative" products to potential markets.

Firms in actually existing innovation systems: knowledge as power

Empirical research on how firms actually behave in spatially defined innovation systems or clusters has raised serious questions about some of the premises that have become conventional or taken for granted in policy-oriented regional development studies (Malmberg and Power 2005; Martin & Sunley 2001). Among the most commonly held premises is the idea that knowledge diffuses among firms that are co-located. It is this commonly available knowledge that enables small firms to introduce innovations and compete in global markets. While the available evidence, including in our cases, appears to support this premise weakly if at all, there has been little analysis of what is actually occurring and how it can be explained. One avenue that has some explanatory potential lies in the analysis of the power relations within firm networks. A closer look at power can give us a different vision of the network as a vehicle for innovation, and change our ideas about what policy interventions might lead to sustainable regional development.

The small firm's role in the innovative network

Policy projects to develop sustainable innovation-based economies typically emphasize the role that small and medium-sized (SME) entrepreneurial firms play as important contributors to the innovative capacity of regions. (Acs *et al.* 1994; Audretsch & Feldman 2003). The idea that small firms contribute to innovative capacity is a compelling one given empirical evidence that small firms "innovate" with greater alacrity than large firms (Hicks & Hegde 2005). Measures such as patent activity, for example, establish a crucial role for small firms in the commercialization of research and development including through university spin-offs.

The ability of small entrepreneurial firms to produce path breaking products and processes also has been linked to their participation in networks of interacting firms (Feldman *et al.* 2005). And in a slight bow to the socio-cultural dimension of economic interactions, studies of regional innovation systems have noted the importance of trust and cooperation in networks and the ways in which positive social relations produce information sharing and knowledge spillovers – the critical factors underpinning sustainable innovation economies (Cooke 2004, 2005). However, empirical studies have found that trust and cooperation among co-located firms are actually quite limited (Angel 2002; Glasmeier 1991; Hendry *et al.* 2000; Lorenzen & Mahnke 2002). Rather, they provide evidence that, even under the most favorable conditions, the relationships among co-located firms, and particularly those in supplier networks, are "close but adversarial" (Mudambi & Helper 1998).

In addition, case studies of evolution in the classic innovative industrial districts of Italy as well other critical case studies raise questions about the *sustainability* of the normative model of cooperative firms (Rutherford & Holmes 2006).Together these accounts suggest a competing paradigm, one in which relations within innovation-based regional economies are infused with conflicts of interest and power relations. This paradigm reflects an understanding of the dynamics of capital accumulation as fundamentally competitive rather than cooperative (Holland 1976). These conflicting relations inherently characterize firm and personal interactions in networks and significantly influence what kind of innovation occurs, when, and how.

To understand how the management of the regional innovation process and the capacity of SMEs to innovate are affected by power relations in inter-firm networks, we need to briefly examine current thinking about the role of power in processes with spatial and territorial dimensions (Allen, 2003, 2004). The proposition that power influences and shapes the allocation of regional capacities and resources among firms has significant implications for analytical approaches to understanding firm networks and for regional innovation policy.

Innovation and power among co-located firms

As we have already suggested, two attributes are considered critical to the ability of co-located firms to become sustainable "regional innovation systems." The first is cooperation within the network of firms. This cooperation promotes a rapid and flexible response to changing and expanding global markets, and the capacity for innovation. Cooperation among co-located firms enables knowledge spillover from the learning and practice of firms in the co-located network. Knowledge spillover and the "untraded interdependencies" (Storper 1997) produced via a cooperative network essentially make the whole greater than the sum of its parts and lead to a sustainable regional innovation system, based in knowledge diffusion. The second attribute is a skilled labor force, which is critical to both innovative capacity and the diffusion of knowledge within and across firms (Malmberg and Power 2005).

Within the policy literature, transnational corporations (TNCs) play a particular and fairly limited role in regional innovation systems. As "leader firms" they connect their fellow network members to global markets, enabling them to grow and expand according to the cooperation paradigm (Porter 1998). According to the standard depiction of their role, TNCs seek competitive advantage by entering specialized regional industries (largely composed of small firms) in order to draw on their innovative capacities and benefit from their skilled labor. The combination of small firm flexibility and innovative capacity with large firm access to global markets enables regions to escape the dominant logic of convergence and price based (or as it is sometimes called) "low road" competition.

Although there are exceptions, the question of power relations has been missing from theories attempting to explain failures in entrepreneurship and innovative capacity. To the extent power relations are recognized, they have been explained with reference to differences among *industries*, leaving a model of cooperation and trust among large and small firms as the dominant paradigm. In its lack of attention to power relations, and emphasis on cooperative relations and "soft infrastructure" the literature on innovative firm networks is afflicted by some of the same theoretical problems as the concept of social capital, which we discussed in the introductory chapter (DeFilippis 2001; Markusen 1999).

Naming as a key process

Power is manifested in a variety of ways including in how products and technologies are named and defined as inside the network or outside. For example, the choice to define a network as a biotechnology network rather than a pharmaceuticals or medical devices network prioritizes technology as the defining characteristic rather than the end products. The same can be said for the industry we studied in Rochester – the photonics industry. The "technology choice" makes the market goals and orientation of the network less visible and supports the background conception that change in markets is primarily driven by changes in technology rather than firm choices.

Another variation on this definitional theme exists in the media entertainment industry where there is a clear power hierarchy defined in terms of gate-keeper control to international markets. So, the independent film industry, although wholly integrated into the production side of the media entertainment industry through a joint workforce, is defined as peripheral to the "core" movie industry – films produced by the "major studios."

Trajectories, contingencies and governance in innovation systems

With the maturing of the literature, a series of case studies have raised questions about whether what have been considered successful regional innovation systems can be sustained over time (Gertler 2003). Among the most important of these case studies is a set which examines the trajectories of firm networks in the industrial districts that inspired the first work on regional innovation and its distinct advantages in global markets (Bianchi 1994; Crevoisier 1999; Piore & Sabel 1983). These studies have found evidence of deterioration in the innovative capacity of firm networks. According to Boschma and Lambooy:

> evidence suggests that in many industrial districts in Italy, there is a tendency for more market concentration (both horizontally and vertically), more market power (embodied by leader firms and business groups), fewer local inter-firm relationships (especially in the case of suppliers and subcontractors), less inter-active and inter-organizational learning, and some signs of institutional lock-in.
>
> (Boschma & Lambooy 2002)

A second set of critical case studies has also recognized limits to innovation by small and medium-sized firms in firm clusters in the auto industry (Belzowski *et al.* 2003; Rutherford & Holmes 2006). These studies, too, focus on change over time and the way in which asymmetries in firm power and differential access to global production and distribution networks affect the innovative potential of firm networks.

From a different quarter, the failed potential of small firm networks embedded in regional innovation systems also has been noted by analysts of what are described as "entrepreneurial regions," which presumably should benefit from trust-based relationships and the sharing of tacit knowledge. The "entrepreneurial regionalists" suggest that what prevents small firms from innovating is the absence of an entrepreneurial ethos, defined as the willingness to take risks in pursuit of big gains and the ability to develop, commodify, and commercialize the outputs of applied research (Audretsch 2004). They attribute the absence of this ethos to market and governance failures that prevent cooperative competition among small firms and inhibit knowledge spillovers.

By contrast with the static orientation of most case studies of regional innovation networks, the case studies that identify problems introduce a dynamic dimension. Their questions and findings are underpinned by a theory of firm path dependency and of industry evolution through the product cycle. This dynamic orientation links these case studies of regional innovation systems to previous case studies of industry evolution and change (Christopherson & Storper 1989; Dicken 1988; Glasmeier 1991, 2000; Markusen 1985; Stone 1973). In so doing, they position questions about regional innovation systems within an existing theoretical tradition in economic geography that examines firms strategies in response to changing market conditions, and government policies, in addition to new technologies.

Secondly the contrarian case studies recognize that hierarchy and power matter in regional innovation systems, that, for example, "leader-firms and other organizations have sometimes become too dominant in the local institutional network" (Boschma & Lambooy 2002) or may be associated with increasing information asymmetries. This concern with power in inter-firm relations is a considerable departure from the conventional literature, which has tended to emphasize trust and cooperation and to imply, at least tacitly, symmetrical relations among firms (Asheim 1992; Asheim & Isaksen 2002). Large firms are conceptualized as intermediaries, linking small firms with global markets. They may be "leader firms" but they depend on the innovative capacity of small firms in the network to provide them with sustainable competitive advantage (Ernst *et al.* 2005; Lorenzen & Mahnke 2002). The dominant research on innovative regional economies emphasized how regional institutions provide the glue that underpins trust and cooperation in firm networks. The case studies that identify the failures of regional innovation networks tend to tie those failures to the governance of the regional firm network and particularly the relationships among the small innovative firms. The failures are attributed either to lack of cooperation among innovative small firms or to industry-specific dynamics that alter relations between large and small firms (Grabher 1993; Rutherford & Holmes 2006).

The "contrarian" case studies suggest that the normative model of cooperative trust relations may, in fact, be the exception and that conflicts and power relations are common in inter-firm networks. Our research builds on these insights but adds a third dimension to the picture – that of territorial governance and particularly sources of control over inputs critical to the innovation system. Another possible explanation for innovative firm network failure, then, is that the unequal power relations and the different strategic agendas of small innovative firms and dominant "flagship" TNCs hinder cooperation, foster information asymmetries, and reduce innovative potential. In other words, the political and socio-dimensions are critical to the functioning of networks of innovation not peripheral. The region is not just a platform for innovative firms to operate in but a created context with its own history of investments, political battles and compromises, and position within a multi-scalar political economy. As the example of the media industry in British Columbia demonstrates, TNCs may be able to use the politically constructed regional context to their advantage and to use the threat of inter-regional competition to make the presence of a regional skill base work to their advantage in producing products and cutting costs.

A realist perspective on regional innovation systems

By contrast with the intra-network focus of much of the literature, the contrarian cases suggest the importance of understanding how small innovative firms and transnational corporations use territorially based governance institutions to leverage regional assets. In the conventional picture, large firms and SMEs appear to operate in parallel universes – the TNC in global markets and the SME in the region. At the same time their interests converge in the arena of innovation, where they play complementary roles (Scott 1992). In reality, the universes of innovative SMEs and TNCs intersect at the scale of the region where they both rely on regional resources to achieve strategic objectives. TNCs and SMEs, however, have considerably different objectives with respect to the content and direction of the innovation process (Harrison 1994a; Storper & Harrison 1991). These differences are magnified if the TNC is publicly traded and subject to pressure for short-term gains (Pike 2005). So, for example, the need for large markets combined with goals of 15 percent annual growth in earnings per share prompt TNCs to focus on new products with large potential growth in the short term (West & DeCastro 2001). The ultimate goal of publicly traded TNCs is less innovation than the achievement of sustainable competitive advantage. To achieve competitive advantage they need to manage the innovation process so that it complements their interests (Ernst *et al.* 2005). As Holmes and Rutherford describe:

> Knowledge development within and between firms within North American automotive OEM supply chains is being shaped mainly by a short term focus on price reduction and the OEMS control of intellectual property. OEM demands for continuous price reductions from suppliers have cascaded down the supply chain and adversely impacted the automotive tooling manufacturers who sit at the bottom of the supplier base.
> (Rutherford & Holmes 2006: 23)

They cite a case study of the automotive industry supply chain on the impact of short-term goals of TNCS, Belzowski *et al.* (2003), who find that "suppliers believe that they transfer more knowledge to larger customers than they receive and too many firms are being forced to focus on short-term cost cutting, at the expense of knowledge-focused production."

In our research on innovation systems in US regions, we also find that large firms that dominate local factor markets and global product markets (TNCs) have different access to resources that are critical to innovation than do small and medium-sized firms (SMEs) in regional agglomerations noted for their innovative potential. Our explanation, however, is based on TNC power relative to governance institutions rather than differences in industrial paths. Because of their influence over labor markets, government policy, and research and development institutions, the needs of large firms tend to take precedence over those of small firms.

One way innovation is managed by TNCs is through control of key resources that support the innovation system. These collective resources include access to key labor segments and to research capacity. In both cases, the interests of small innovative firms and those of TNCs attempting to achieve sustainable competitive advantage in global markets intersect and clash at the regional scale.

Competition over key resources: research capacity

Our research in Rochester pointed to an important source of inequality and asymmetry between large TNCs and innovative SMEs: access to research infrastructure. Again, regional innovation systems are defined and organized to serve particular interests.. For example, in industry cases we have examined, TNCs want research institutes supported by universities or public funds to take on specific tasks in the development process. Because they cannot directly control university-sponsored research, they prefer that universities focus on generic technology, giving the TNC direction as to what research avenues are likely to be more profitable. They can then rely on in-house or captured research institutes to do the research that will result in commercial products and processes (Ernst *et al.* 2005). Increasingly those "captured research institutes" have emerged as partially publicly financed innovation centers or (as they are called in New York State) "centers of excellence."

Universities, however, play a conflicted role in the regional innovation process. University sponsored innovation centers are engaged in developing innovative products and processes that can be patented and sold or licensed to provide the university with a stream of income. The typical process of small firm buy-outs by large firms underscores the very different firm strategies employed by small and large firms. Universities interested in innovative university-led economic development, have found that transnational corporations are interested in "embedded labs" in part because they provide the TNCs with access to the emerging entrepreneurs and their ideas before they spin off into competitors but after a technology is developed. This environment mimics the historic research and development relationship between small and large firms developed in the optics and imaging industry in Rochester.

Research on the innovation center agenda in the field of biotechnology indicates that spin-offs to SMEs are rare and that commercialization is limited to those products of interest to large transnational firms, such as those in pharmaceuticals (Gertler and Levitte 2005; Kenney & Patton 2005). The large firm research investors may not be located in the region of the innovation center so investment in innovation only occasionally contributes to the development of a dynamic regional agglomeration or to the broader regional economy. In the context of the goals of these centers, small innovative firms are a means to an end, rather than a resource whose potential contributions to regional innovation need to be fostered.

In another example of the conflict between regional innovation capacity and TNC agendas, TNCs are opposed to "the over involvement of universities in downstream product development activities" which occurs as a consequence of the university goal to increase capital flow from equity holdings in start-up firms and patents and licensing agreements (Feller 1999). The small firms that universities spin-off, however, lack in-house commercialization capacity (connection to global markets) and need assistance in converting generic research into commercial properties (ibid.). Since universities are critical venues for this kind of support, TNC emphasis on pre-commercial research focuses on their needs and directs attention away from downstream applications that benefit smaller firms.

There are other examples of divergent interests between large transnational firms and the SMEs who both supply them and could potentially compete with them on the basis of innovative capacity. One prominent and relevant example is state-supported "innovation centers." In the US, many states, along with their federal partners, have invested significant economic development resources in these industry- or technology-specific centers aimed at research, training, and commercialization. A stated goal of these centers is to promote regional innovation capacity by nurturing nascent entrepreneurs. Publicly subsidized research centers are presented as a lynchpin of state innovation policies (Bozeman 2000; Bozeman & Boardman 2004; Coursey & Bozeman 1992).

The agenda in these centers, however, is heavily weighted toward the needs of the large firms. For example, many centers are developed around explicit partnerships among universities, large firms, and state and local government. In the cases we have examined, the centers are managed by staff seconded to the innovation center from transnational firms in the region and their advisory boards are dominated by large firm representatives. In order to benefit from the resources of the center, small innovative firms must be willing to compromise their independence and control of their intellectual property, allowing the large firms to learn about their innovative activities. Information flow tends to be upward rather than diffused or horizontal.

The composition of the Board of Directors for the New York Infotonics (Information and Photonics) Center of Excellence provides some insights into the differences in power and influence of transnational corporations and SMEs over regional innovation assets, including valuable information. The Infotonics Center lists ten Board members, seven of whom currently work for Eastman

Kodak, Corning, or Xerox, the regional TNCs. The Board is rounded out by two directors of university research centers and one small firm representative. The CEO is an Eastman Kodak retiree. While innovation centers have become popular sites for regional economic development investment, their governance remains dominated by large corporate and institutional interests, calling into question their role as engines for small firm growth and new product innovation and commercialization.

Far from providing a regional entrepreneurial ethos, these centers seem to provide spaces for large firms to observe their small firm rivals. In some cases, large firms negotiate a "right of first refusal" for innovations developed under the center's umbrella as a condition of their participation. These deals undercut the role centers could play in nurturing a regional entrepreneurial ethos.

TNC definition and control over the agenda of the innovation system is one reason why regional innovation systems fail to equate with sustainable regional development. A second reason is control over another resource critical to regional development in a knowledge-based economy, that of medium and high skilled labor.

Competition over a key resource: skilled labor

One factor that is taken for granted in the regional innovation literature is the critical role of highly skilled labor. Our research in regions with specialized industries, including those in advanced manufacturing (Christopherson *et al.* 2007) indicates that, because firms in high tech and advanced manufacturing industries are engaged in product commercialization and prototype construction as well as research-based innovation and small batch specialized production, labor force needs are more complex than they have been portrayed. Although science and engineering workers are regularly considered a locational asset in attracting and retaining so called knowledge or innovation-oriented firms, the labor market needs of the industry include workers with a range of skill levels (Florida, 2002a, 2002b; Gertler & Wolfe 2002). A US National Association of Manufacturers survey of 800 firms in 2005 showed that 80 percent of firms said they were experiencing a shortage of "qualified" workers. While they describe their need as one for skilled workers, when queried about specifics they described their skill needs in terms of manufacuring-based occupational skills, such as welding, soldering, and machine tooling – skills that do not require advanced degrees (Hagenbaugh 2006).

Perhaps, ironically, a skill shortage is occurring in those innovative firms that have moved from routine manufacturing using a semi-skilled workforce to advanced manufacturing, drawing on a smaller workforce of medium and high- skilled workers. So according to a report from the Federal Reserve Bank of New York, employment in high-skilled manufacturing jobs rose by 37 percent from 1983 to 2002 while low-skilled job dropped by 25 percent (Hagenbaugh 2006). Behind this up-skilling trend is continued pressure on small and large manufacturing firms to adopt "lean production" methods and to do more with less.

In the Rochester regional labor market, SMEs, including entrepreneurial innovative firms, find themselves in direct competition with transnational firms for workers with advanced manufacturing skills. This problem was consistently reiterated in interviews with the CEOs of SMEs, with public officials, and with the Rochester Industrial Development Agency, charged with assisting the expanding group of innovative photonics firms. During our research in the early 2000s, the supply of medium-skilled labor, particularly operatives with machining skills, was extremely limited. In part, this resulted from the power of large firms in setting the prevailing wage in the region. Prospective employees looking at the Rochester region would decide against moving there because wages were consistently below the national average for their skills (Clark, 2004; Pendall *et al.* 2004). Our interviews indicated that TNCs actively lobbied public officials to prevent the entry into the labor market of competitor transnational firms that would raise the prevailing wage rate by competing for medium-skilled workers. The evidence from Rochester has parallels in other industries and in other regions.

In another case, The Georgia Manufacturing Survey, a statewide survey of manufacturing firms conducted about every three years providing an usually detailed time series of conditions within firms showed a similar pattern of competition over medium-skilled workers. The 2005 survey included 648 respondents, 80 percent of whom were manufacturers with between 10 and 100 employees. The survey is heavily geared toward questions about technology, innovation, and research. However, in the questions about what firms need in order to grow, expand, and build competitiveness, there were fifteen multiple choice options including eight that were relevant to labor skills and work. Only five options out of the fifteen categories received responses from more than 20 percent of the firms. Of those five, four were in areas of workforce or skill needs, with basic workforce skills ranking second and work process and flow ranking the highest. A similar question about labor market skills elicited results that are consistent with the demand for medium-skilled labor seen in the Rochester survey.

Unexpectedly, perhaps, competition is not as severe at the high end of the skill spectrum. Small innovative firms attract entrepreneurial skilled labor because they offer more interesting jobs and the alternative compensation that makes them competitive with corporate employment (e.g. stock options, co-ownership, leadership positions). Our research indicates that engineers and other high skilled workers are willing to take pay reductions in order to work in smaller, innovative firms. This option may not be open to them, however because policies promoted by TNCs prevent skilled workers from moving among firms. "Non-compete agreements" essentially stop the transfer of technology and limit innovation by preventing skilled workers from moving among firms or establishing new enterprises in their area of expertise.

Because TNCs operate in a global labor market, able to attract engineers and scientists from all over the world, they could provide the regional skill base for technological innovation in existing and new firms. Instead, covenants not to compete protect the recruitment and training investment of TNCs by limiting the mobility of workers. This inability is not uniform however. It is notable that the state of California, which is noted for its innovative "culture," does not have non-compete agreements (Gilson 1999; Saxenian 1994; Stone 2004).

Access to knowledge and labor skills are critical to the functioning of innovation systems. Evidence that these collective resources are shaped and directed to meet the needs of some firms over others suggests, however, that we need to take a more skeptical look at the ability of innovation systems to produce regional economic development.

As a model for economic development, the regional innovation system has an unfortunate flaw. It is not clear how innovation capacity creates jobs. While the innovation stage of production may have the highest added value, it also requires the smallest quantity of workers. Further, the innovation stage of production has high capital costs in terms of research and development, and periodic, rather than consistent, income streams coming from patent sales, initial public offerings, and capital investment rather than a steady stream of product sales. With the uncertainty of this model comes the demand for flexibility and the need to effectively manage and redistribute risk.

As the case of optics and imaging demonstrates, innovation in digital photography, while producing a new market linked directly to the computer and media, also negatively affected the existing market for conventional photographic equipment. Some skilled jobs were lost just as others were created. So, the proposition that regionally based innovation results in regional production is a vast over-simplification of complex economic processes and production decisions. The assumption that regional innovation or, for that matter, the use of high-skilled workers, will, a priori, result in job growth is, at best, optimistic. Technological change produces a wide range of regional economic impacts – employment growth is only one possible scenario.

In this chapter, we have examined how innovation systems, composed of firm networks and the institutions which support them, are rhetorically linked to regional economic development and thus to regional job expansion and an improved standard of living for regional residents.

Our own research, and a considerable body of empirical work, indicates that this link is tenuous at best, particularly within the Anglo-American "variety of capitalism." When firms are under severe pressure to produce short term gains for investors and have a singular amount of influence on the governance institutions within which they operate, they are inclined to distribute the risks and costs of innovation to regions and the workforce. Firms tend to capture the rewards of innovation for management and the stockholders rather than allowing them to be reinvested in the regional economy. Because of these "perverse incentives," firms frequently face diseconomies in the region, in the form of labor force shortages that drive them to look for other regions to locate various parts of the production process.

In both the regions we have studied, Los Angeles and Rochester, dominant TNCs have relocated production processes in regions outside the U.S. where they have access to less expensive skilled workforces. While they have retained "research and development" activities in Los Angeles and Rochester, it is not clear that the production that results from innovations in products or processes will occur in these "home" regions. While there is a greater chance that new ideas developed and commercialized by smaller firms founded in the region will result in local employment, our research suggests that these firms are disadvantaged when it comes to tapping the resources they need to commercialize their products and grow in place.

In recognizing the real life distance between innovation and regional economic development, we share the skepticism of John Lovering (1999) about the "hype" surrounding regional innovation systems and concerns about the considerable public investment in "clusters" or regional innovation systems on the grounds that they will enable the region to slip through the net of cost-driven competition.

In the next chapter, we examine how some of the limitations of the regional innovation system have been addressed by putting the region, its quality of life and the collective resources available to its citizens, at the center of the question of regional competitiveness.

7   The learning region disconnect

The concept of "the learning region" has emerged from a history of attempts to name and describe the characteristics of places that respond successfully to the competitive challenges presented by a knowledge-based economy. Michael Piore outlined the components that define successful regional innovation systems and industrial districts in 1990, laying the groundwork for an extension of the regional innovation system to a broader conception of regional attributes associated with innovative economies (Amin 1999; Scott 1992, 1998; Sengenberger et al. 1990; Storper 2002):

> Thus far, the literature seems basically to have identified a list of factors that are critical to success. The standard lists include: 1) a major research university, 2) an academic tradition, or ethos, which encourages researchers to engage in practical activities and which is not hostile to linkage between the academic and business community, 3) venture capital or, more precisely, a local financial community with both the resources and the willingness to provide funds for start-up enterprises; and 4) a local entrepreneurial tradition and a reservoir of expertise on the management of start-up business. The attempts to create new [high tech] regions have essentially tried to create the institutions on the list.
>
> (Piore 1990: 299)

In this chapter we examine the expansion of the regional innovation systems framework to encompass the "learning region." As an extension of firm-based innovation models, "the learning region" is centered around the interests of firm networks and an elite population of knowledge producers and knowledge users (Cooke and Piccaluga 2004; Florida 2002b). The overall economic rationale remains the same – to enable the region to compete with other regions in what is presented as a zero sum game. Some regions will be winners and others will be losers. Whether a region "wins" depends on the ability of key players in the regional economy to support the international competitiveness of "its" firms (Lagendijk and Oinas 2005)

The learning region, however, broadens the list of what regions need to provide in order to compete successfully in the global knowledge economy. The emphasis in the learning region is on place-specific resources, attributes, and institutions, and on the ways in which they can be deployed to meet the needs of firms located in the region but competing in a global economy. While much of the literature continues to address the broader regional context for innovation systems in abstract terms, such as milieu or ethos, the learning region has been "operationalized" to focus on: 1) a central role for institutions of higher education in providing a range of research and development services, the work-force skills required by firms, and an environment that will ensure the success of firms engaged in the kinds of production that requires "brainpower", and 2) local (and where it exists, regional) government support of firm competitiveness through coordination of resources important to the innovation network, and direct and indirect subsidies to internationally competitive firms.

Proponents of the learning region assert that investments should be made in the infrastructure that supports the regionally located innovative firms and that both institutional and governmental resources should be directed to drive the success of the innovation system. By talking about real public monies and real institutions, they highlight questions that are obscured in the firm-network oriented innovations system discussion. In essence, they move the discussion from abstract academic arguments into the arena of policy-making and, thus, into the realm of power and political representation (Cooke & Morgan 1998; Morgan 1997).

The move to make the learning region "real" through policy initiatives was a critical turning point in the literature on regional innovation systems and learning regions. It focused attention on the rationale for investing collective resources in actually existing innovation systems and raised a key question: Does the use of collective resources to support innovation systems benefit the wider regional economy and workforce? The turn to policy also raised a set of governance issues that moved well beyond governance of the firm network: Who is included in the regional innovation system project? Who decides how place-based collective resources are going to be used? Who sets the priorities? On what basis do they make their decisions?

Because turning ideas into policy prescriptions raised questions about *who* in the region was making decisions about the use of collective resources and *to what ends*, it also connected what had been a limited discussion focused on intra-regional firm networks to a broader discussion about the meaning of "success" in regional economies. In addition, because the learning region policy prescriptions were extended to regions characterized by disinvestment and economic stagnation, they raised older questions about the sources of uneven development.

These questions arose from various quarters and concerned different scales of decision-making and governance. They first emerged in concerns about the region as a "privileged" locus of economic action in the global economy.

The origins of skepticism about regional innovation policy

Florida and Kenney (1990) were among the first skeptics to raise questions about regional development policy organized specifically around innovation systems in their critique of the Silicon Valley model, entitled, "Silicon Valley and Route 128 won't save us." They argued that high-technology industries in the U S. do not operate in the same way as those in flexibly specialized industrial districts in Europe and pointed to the way in which the rules that govern investment in and across firms significantly alter the incentives that shape firm behavior. In practice, and as we described in Chapter 6, large firms in the US squeeze their suppliers rather than collaborating with them. Firms sue each other over intellectual property disputes rather than sharing innovative ideas across firm and institutional boundaries. Firms limit technology transfer and the success of entrepreneurial spin-offs by lobbying for and implementing non-compete agreements.

Using evidence from firms operating in Silicon Valley and Route 128, Florida and Kenney demonstrated that while US firms may locate within geographic proximity to one another, they do not reap the advantages that proximity enables in the industrial district paradigm. Instead they act on incentives to redistribute risk to captive suppliers and emphasize market dominance and corporate profits over the production of market-altering innovative products (Florida & Kenney 1990). Florida and Kenney's early broadside against "the Silicon Valley Model" suggested that governance institutions and structures beyond the regional scale affect how firms interact with one another and with their regional work-force. The skepticism around the "Silicon Valley model" was deepened by later evidence that, in addition to innovation, the Santa Clara valley regional innovation system also produced intensified inequality and diseconomies of scale (Benner 2003; Carnoy *et al.* 1997).

Concerns about whether this quintessential US model could be generalized have been reinforced (and its theoretical underpinnings deepened) by research on the "varieties of capitalism." Although open to critique about the direction and speed of change in differently configured capitalist economies, the varieties of capitalism approach has been an important corrective to simple economic explanations of how a global economy is emerging and of the bases for competition in that economy. As we noted in Chapter 2, research from a perspective that recognizes and takes political-economic institutions seriously points to the importance of political power in constructing differently configured market governance systems. And, although the varieties of capitalism approach focuses on national systems, it raises serious questions about whether firms in a region can operate under different incentives than those that structure the broader market governance system within which they emerged. Thus, by extension, an understanding of market governance from a political perspective suggests that there are limits to explanations based exclusively on intra-regional processes.

From this vantage point, we can look at the US innovation environment as having particular strengths and weaknesses and at the region and regional resources as playing a distinctive role in that system.[1]

We can see elements of "systemness" in the comparative national patterns of support for research and development produced by the US National Science Foundation (2004). Although dated, these statistics indicate that contemporary US research and development is heavily oriented toward defense industries by comparison with other OECD countries, particularly Japan. There has been declining federal support in the US for non-defense research and development and a considerable uptake in university-based research activities funded by industry sources. What these trends suggest is a rationale for moving to sub-national state sources for funding research and development to replace federal funding and, as we will lay out in the next section, a need to emphasize the regional character of what are, in reality, major international research centers, whose industrial innovation research is primarily oriented to and supported by TNCs.

A second major distinguishing feature of the US innovation environment is a historically decentralized policy-making environment. Variation in regulation at the sub-national state level distinguishes the US from other advanced economies (Befort 2003). This decentralization works against the kind of national innovation systems that characterize innovative economies such as Sweden, Finland, Japan, and South Korea (based in centralized power, or social partner coordination). It encourages inter-jurisdictional competition via subsidies to individual firms, including for their research and development and human resource training expenditures. Ultimately, industrial policy in the U.S is dominated by ad hoc policies aimed at individual firms (Markusen 1991). US politicians rail against the idea that the state could adopt industrial policies that "pick winners." In reality their problem with a national innovation system is that it limits opportunities to obtain campaign contributions from individual firms.

In addition, the national innovation system is largely determined by the interests of large transnational corporations because of their ability to shape policy at the federal and state level and the political fragmentation and weakness of other interests, including labor and small and medium-sized enterprises. As we described in Chapter 6, the US version of the innovation system is not neutral. Resources are used in particular ways and controlled by the more powerful players.

It is in this political-economic context that the region emerges as *the* locus for global competition, under the rubric of "the learning region."

The regional fix: its uses as a narrative about contemporary capitalism

To understand why the regional fix, particularly in the form of "the learning region," is so compelling in the US, and why it has taken a particular form, it is useful to return to a widely accepted story explaining "globalization" and its spatial implications.

The rise of the region as *the* locus of economic action in the global economy begins from a conception of the role of the state as an intermediary between customers and markets. This conception is associated principally with Ohmae (1995), who describes the nation-state as an unnatural, dysfunctional unit for organizing economic activity in a borderless world and the "region state" as the appropriate unit of analysis. According to Ohmae, the region becomes the central unit of activity in the global, knowledge economy because, by comparison with the nation-state, "the region" is a space of coherent economic interests and integrated transactions.

This spatial theory underpins a particular interpretation of the direction and end point of global market integration, the "convergence" model. As Berger (2006) describes: "The breakdown or the negotiated surrender of national controls over the flows of capital, goods and services across borders means that producers everywhere find themselves in competition." With the inevitable state "surrender" via deregulation, market convergence "shrinks the resources under national control for shaping economic and social outcomes and the legitimacy of national cultures and institutions."

What remains is the natural economic zone – the region – which emerges with the decline of nation-state authority and legitimacy. From this seed comes the idea that regions and the firms that operate within them exist apart from politically constructed rules, such as those regulating property rights, competition, or labor relations, which are established at the national scale and depend on state enforcement. As critics of the new regionalism have noted, however, the absence of a theory that embeds the firm and the region in broader scales of market governance limits the ability of the regional project to connect with other narratives about the way the global economy is taking shape (Lovering 1999).

The idea of regions as natural economic units that have emerged with the end of nation-state power is, however, a compelling idea, particularly when combined with the historical US institutional construction of inter-jurisdictional competition and cultural attachment to a conception of the market economy built around zero-sum competition between winners and losers. While the theory of the region as a natural economic unit provides a useful rationale, the program connected with region-centric globalization theory has its roots in real political projects, specifically those to liberalize trade, deregulate industries, and devolve responsibility for basic social protections to the local scale (Brenner 2004). Policy projects, initiated at the supposedly powerless national scale, put pressure on local policy-makers and institutions to accommodate the needs of internationally competitive firms if they are to remain viable. From the perspective of the firm, strategies to achieve sustainable competitive advantage can be pursued at multiple scales. If the critical human and physical infrastructure required to support the firm's needs is not being provided by the nation-state, the region becomes a target of opportunity and a critical scale for constructing competitive advantage. For the TNC, the learning region is a vehicle to marshall resources at the regional scale in the interest of global competitive advantage.

The use of collective resources in the interest of regional innovation

A term frequently used in the literature explaining the workings of the learning region is "collective." In a parodied form, "Collective resources are applied in such a way as to increase collective capacity." Despite the persistent use of the terminology, however, the process through which a set of collective assets can be developed and used for the purpose of building innovation capacity is not straightforward. Collective assets are like social capital – a nice idea in the abstract but very complicated in the real world. In the US, the idea of "a commons" whether tangible like a research center or theoretical like an innovative milieu, directly conflicts with a culture of competitiveness that privileges property rights – intellectual property, proprietary knowledge, patents – over collaboration. In addition, as we laid out in Chapter 6 and in our case studies, the ability to use collective assets is a political question, with some actors having more influence than others.

Although not stated explicitly, there is a tacit understanding that collective goods are available for use in economic development, and that the use of collective goods to serve the innovation system will rebound to the benefit of the polity and the broader citizenry. Evidence on actually existing innovation systems in the US suggests, however, that there is a gulf between the promise of the learning region and its ability to deliver on that promise. At the very least, as the pressure increases to use public monies to support firms with connections to global markets, the efficacy of innovation-based public investment raises questions concerning relative value and opportunity costs (Amin 1994).

The promise is based on two assumptions. The first is that the adoption of learning region policies will increase the international competitiveness of firms located in the region and enable the region to attract investment and avoid the trap of cost competition. As our case study of the media entertainment industry illustrates, however, TNCs can construct cost competition even among high-skilled, regionalized labor pools. In addition, as we laid out in the previous chapter, the power of TNCs in potentially innovative networks may inhibit the innovative capacity of small firms and drive them toward cost competition. One result is that regional policy aiding the competitiveness of the "big players" in the international economy may only marginally benefit the region. If local and state governments are providing subsidies to the TNC, in the form of tax abatements, the benefits to the region are further decreased.

Since TNC ability to pit even high-skilled regions against one another and to dominate firm networks derives from macro-economic policy regimes, the routes to address the unequal bargaining power of the region and the cost competition trap lie *outside* the direct control of intra-regional actors, at the sub-national state and national scale.

The second assumption is that the benefits of regional investment in innovation systems are directly linked to a healthy, sustainable economy. As the Silicon Valley example demonstrates, even what are widely acknowledged as highly successful regional innovation systems can be characterized by high levels of inequality and dysfunctionality associated with diseconomies of scale (Benner 2003). As the income of the privileged few in the knowledge economy increases, the benefits of learning region "investments" rarely reach the working class.

In addition, without the kind of redistributive policies and regional economic development programs that extend the opportunities of the growth in knowledge-intensive sectors to the citizenry as a whole, the viability of the regional innovation system is threatened. In his study of the knowledge economy in so-called "underperforming" regions Sokol (2003) effectively jettisons the links that have been assumed between regional innovation systems and widely shared regional prosperity. "By assuming a causal connection between the knowledge economy and wealth, the learning region basically replicates the central flaw of the knowledge economy concept."

Given these broadly defined limits to the learning region as a policy program for regional economic development, we can take a closer look at how one critical collective resource, institutions of higher education, is used in US style innovation systems.

Universities as "economic engines"

The adoption of the learning region concept in the US has focused on a key institution, the university, and a central project, public–private initiatives to use higher education institutions to serve the needs of regional innovation systems. Key components within this project are university-sponsored programs to aid large firms looking for more flexible ways to conduct research, and to provide start-up firms with skills and sheltered environments within which to grow. While these activities are not, in and of themselves, inconsonant with the university mission, they are secondary to the central mission – the development of human knowledge and expression. To the extent that the competitiveness-oriented goals behind regional innovation systems interfere with or shift resources away from the central mission, they undermine the university as a knowledge-creating institution. The most compelling regional success stories, including that of Silicon Valley, demonstrate that the preeminent contribution of universities lies in the quality of university graduates (Benneworth 2006).

What is new in the current debate is the intensified pressure placed on universities to serve as "economic engines," supplanting government as the institutional agent responsible for nurturing a healthy economy. In part, this newly created "job" for universities is a consequence of devolution and the decreased capacities of the local or regional state. As our discussion in Chapter 6 pointed out, universities, their research centers and government labs are key components of innovation systems. In addition, universities have been identified as critical to regional competitiveness and economic development because of their financial assets and skill development capacities (Harloe & Perry 2004; Quintas *et al.* 1992; Saxenian 1994; Wolfe & Holbrook 2000).

The evidence concerning the positive regional impact of state investments targeted specifically to innovative capacity, including in universities, is, in fact, mixed; the results linked to industry characteristics and individual firm strategies (Egeln *et al.* 2004; Feldman & Desrochers 2003).

US universities have a complicated agenda when it comes to technology transfer and innovation. To some extent, that agenda is driven by the limited portion of university budgets that is covered by tuition costs, and the increased demands on universities for student services and research infrastructure. A top ranked university in the United States must attract research dollars, and in order to attract research dollars must have the up-to-date infrastructure to support research. This spiral has driven universities to support more industry-sponsored research and to look to the sub-national state for support to provide facilities and staff. Adding to the competition and the cost spiral is "the star system," which provides extraordinary rewards to faculty who receive the most public recognition, and the need to support faculty who bring in substantial grant monies to the institution.[2]

While costs have spiraled upward, public support for higher education has declined because higher education investments are seen as discretionary when compared to other demands on the state coffers, for example health care. The combination of increased costs, opposition to rising tuition, and demand coming from industry to use university research facilities and resources has driven university officials to emphasize service research for major firms. They have also ramped up divisions that derive income from technology transfer, licensing, and patents. In short, research universities have begun to emphasize their innovative capacity, not as a route to enhance the strength of the regional economy in which they reside, but as a way out of their cost crunch. They have become entrepreneurial in selling both research services and innovations to increase and secure their financial condition.

Studies have demonstrated that the relative cost of performing these research functions at the university has increased while the indirect cost recovery allowed by the federal government has declined. For example, Cornell University's indirect cost recovery rates for the endowed side of the university dropped by 20 percentage points between 1991 and 1997. The fraction of total research and development expenditures contributed by universities has risen from a low of a little over 10 percent in 1973 to almost 21 percent in 2000 (Ehrenberg *et al.* 2003). Thus universities are beginning to subsidize funded research functions through their operating budgets, not replace declining state and federal funding with new outside funding streams. Again, a national strategy for investment in research and development capacity would seek to mitigate these negative externalities that catch universities between their outreach and educational missions. Whether this evidence is indicative of future trends is debatable, however, initial results point to a need for rethinking the model.

For many reasons, including prestige, money, and mission, universities have not resisted the research and development investments that have come their way. Universities fit the criterion of a regional economic development policy struggling in the United States: they are fixed in place. And, to an increasing degree, universities are becoming innovators themselves, taking on much of the research and development risks that once were the purview of large firms. Ultimately the role of the university in developing regional research capacity is another example of the shifting of firm boundaries from internal to external activities.

What is not clear is whether universities are appropriate vehicles for the types of economic development activities that are now emerging as innovation becomes the metric of development success rather than jobs. The questions of governance and accountability, equity, job creation, and university mission all need to be addressed as the negative externalities of the university-oriented approach mushroom: gentrification, higher costs to the university, and ownership conflicts over intellectual property.

What universities can and cannot do

If we examine the potential contributions of the university in regional economic development relative to the overall needs of economies attempting to foster long-term sustainable economic development, we see that universities can play only a limited role. Without a regional capacity to absorb technological innovations and support new firms, university innovations will be developed and commercialized by firms outside the region. And without a commitment to the development of a broadly skilled workforce, apart from a small group of graduate engineers and computer scientists, the region's innovative capacity will falter.

In established innovation centers, located in the megalopolitan regions on the US coasts, there is a great likelihood that spin-off firms or the relatively small amount of job creation associated with technology transfer will be retained in the region (broadly speaking) of the university. Research on the commercialization of patents produced in universities outside these centers shows, however, that while universities in the newly constructed periphery may be producing innovations, production is most likely to migrate to the coastal centers (Bania *et al.* 1993). Although the failure to attract innovative production firms typically is attributed to the absence of regional leadership, entrepreneurial spirit, or regional ineptitude, or, in Florida's words, "the region's receivers are turned off or not working properly" (Florida 2005), there are specific policy-based explanations for why coastal locations have become preferred by firms seeking economies of scale, a large skilled labor pool, and venture capital. Industry deregulation in the 1980s and 1990s in transportation, energy, and communication raised costs in regions outside the coastal megalopolitan regions and initiated a process of disinvestment. It is the consequences of this disinvestment that drive firms away from university-based innovation centers in the periphery to the coastal regions. This spatial reallocation of resources was neither natural nor accidental but the unintended consequence of macro-economic policies adopted at the national scale to make "national champion" TNCs more competitive in the global economy.

While universities can contribute to the innovative potential of a region, they cannot make up for extra-regional policy decisions that have produced: 1) an inadequate transportation system; 2) expensive access to global markets; 3) poor public education; and 4) inadequate access to broadband internet. The gap between university innovation and production location is a question of macro-economic policy and its effects on regional capacity not of university resources and intent. If the region surrounding the university does not have the capacity – in terms of management skills, labor force, market access, reasonably priced public services, or venture capital – to absorb university-produced innovations, then those innovations are likely to end up far from the point of their origination.

Measures of economic development success applied to universities also have been expanded to include the number and skills of knowledge workers that are retained in the region adjacent to the university; and the number of jobs and firms created where the university is located. In this respect, too, universities are limited in what they can contribute. Although there is a strong demand for college-educated workers, a "knowledge economy" actually requires a generally skilled workforce in order to thrive. For example, studies of advanced manufacturing industries in the US show a serious labor shortage among craft workers with skills in soldering, welding, and machine tooling (National Association of Manufacturers 2005).

As our study of the Rochester photonics industry and its labor market demonstrate, these middle-skilled craft workers are sought after by both large TNCs and smaller innovative firms. These craft workers do not fall under the descriptors used for "the creative class." They do not have advanced degrees but rather are the product of state and local investment in high quality basic education. The current US labor shortage for middle skilled manufacturing workers in fact reflects the problems that result from exclusive focus on the creative class and neglect of the working class. It should come as no surprise that economies, such as those in Germany, that have better redistributive mechanisms and a higher average standard of living also do better in producing this critical base of skilled production workers. By contrast, a population without skills reflects a long-term failure to invest in education and health care, the basics underlying the "virtuous circle" that produces healthy sustainable regional economies (Bartik 1991). This lack of long-term investment cannot be solved by the quick fix of university-trained personnel.

Finally, universities with a public mission, such as "land grant" universities in the US, find themselves in a bind – caught between their core mission and the demands that they serve a much wider public policy agenda while at the same time dealing with a decreasing funding stream. Of course this conflict is much less pronounced for private institutions and so, even historically, publicly supported institutions in the US are considering "going private." Public institutions that retain a mission to contribute to the collective good find themselves, perhaps ironically, in a situation analogous to that of the local public sector – with multiple demands for services and an uncertain and decreasing flow of reliable capital support.

The learning region – a theoretical "lock-in"?

In the now substantial literature that extends the concept of regional innovation systems to the learning region, earlier frameworks that associated regional competitiveness with innovative firm networks have developed substantially. The learning region concept opened the eyes of researchers to a broader set of regional conditions and institutions that affect the ability of regional innovation systems to function. In response to this broader and more open framework, questions have been raised about the framework itself, with researchers pulling back and qualifying their interpretations about the relationship between regional innovation systems and healthy regional economies. These qualifications have become particularly apparent when the research agenda has extended beyond the successful "hotspots," to the wide range of "hard cases," places where the learning region is not present either in a set of institutions or a set of policy-driving ideas.

The study of regional failure has been important because it has led researchers to examine how political institutions and policy decisions external to the region affect its ability to emulate "hotspot" models. This constitutes an acknowledgement that institutions and political economic regulatory frameworks beyond the regional scale exert influence on regions (Braczyk *et al.* 1998). Regional failure has been couched in terms that replicate the framework first presented by Ohmae (1995), who suggested that the region was a natural economic unit in the global economy because of its "coherence." The analysts of regions that fail to thrive find this failure in "incoherent" regulatory frameworks. "So, . . . the internal coherence and compatibility of the local order must be taken into consideration so as to arrive at a qualified judgment of a given regional innovation system" (Ohmae 1995). Our case studies and analysis of power in firm networks tells us that a concept of regional failure based on incoherence doesn't capture the political processes and power relations that shape regional destinies.

As more questions have been raised and qualifications registered, the learning region framework itself seems to demonstrate problems of institutional lock-in. Its proponents routinely adapt in marginal ways to theoretical or explanatory challenges while failing to move to a new paradigm. The emergence of a new paradigm will require serious attention to questions about "the region" and the processes of spatial transformation that have heretofore eluded attention.

As a prelude to addressing policy directions, we raise three sets of questions that need to be addressed to move to an understanding of the contemporary regional question.

The missing dimension of labor and the regional labor market

The most obvious definition of the region, based in the regional labor market, is missing from the learning region literature. As we noted early in this book, recognizing the workforce as a critical component of sustainable regional economic development opens up questions about the broader conditions that affect the development of skills and innovation, questions that go far beyond those contained in arguments about the creative class.

Who is acting in and on regions? Who across regions?

The idea that "regions" act, make decisions, and carry out policies obscures the power relations among firms and between labor and capital. This conflation of complex and competing interests within a single concept of "the region" constrains an analysis of the conflicts underlying regional policy and the political process of building coalitions to promote particular visions of regional success.

The concept that the region is an actor and responsible for its own destiny is used in a narrative that blames the regional victim. As Sokol (2005: 75) describes, the regional problem is conceptualized as an outcome of differentiated levels of learning and innovation. Less favored regions are under-performing due to their inferior innovation (or learning) capability.

This last is key because it suggests a direction for understanding contemporary uneven development and the use of the region as a discourse to explain away spatial inequality and the patterns of investment and disinvestment that have produced it.

What are regions competing over? Who is served by competition? Why do we understand inter-regional competition as a natural (rather than constructed) phenomenon?

In the US, Atlanta and Miami are locked in a battle to determine which city-region will become the gateway to Latin America as the Free Trade Area of the Americas moves forward. This competition involves luring the North American or US headquarters of Latin American transnational corporations. This competition between regions is emblematic of the multi-scalar nature of inter-regional competition in the global economy and the slippery definition of what is regional – the city-region or the multi-national trade zone.

The extension of competition to the instrumental use of collective resources

Governance and accountability remain serious issues for innovation-based policies and for regions. While innovation strategies aimed at regional institutions rather than firms avoid the classic problem of "picking winners and losers" among firms in the same industry, they do not address the question of governance that permeates the regional discussion. While avoiding the intra-regional economic development problems, institutional innovation strategies retain the problems of inter-regional competition (Malecki 2005). As with firm specific subsidies, government funded university research has the potential to mirror the practice of the zero-sum game for localities as universities and regions learn to bid against each other for innovation dollars. From an equity standpoint, the burden of industrial policy is better shared at the national level, mitigating the motivation of places to bid against each other (Lovering 1999; Malecki 2005; Markusen 1991). A national industrial policy ameliorates many of the competition problems between states and regions for university research dollars. Perhaps for this reason, most industrialized countries organize state-funded investment for innovation in a less ad hoc manner (Wolfe & Holbrook 2000).

What does this suggest for policy directions? In the next chapter we raise the possibility that the regional question, including the contemporary processes constructing uneven development, needs to be addressed not only in the region but at a governance scale that can undercut the wasteful inefficiencies produced by inter-regional competition and redistribute the resources necessary to extend the capacity for learning and knowledge creation beyond a small portion of the population.

Theoretically, the disconnect between the capacities and economic health of the broader labor-force based region and the strategies advocated for "the learning region" needs to be more closely scrutinized, not least because it is associated with trends toward social inequality and segregation.

Ultimately, the learning region disconnect is attributable to the incoherence of "the region" in the US context, and an unproven connection between regional innovation systems and the collective good and regional prosperity. At the heart of this disconnect is the rationale presented for supporting the deployment of collective resources in service of the innovation system and the reality of how firms view and use regionally fixed resources. As our case studies describe, the TNCs see regions not as self-contained economies but as sets of opportunity structures – for tapping talent pools, reducing costs, obtaining inputs, including knowledge inputs. As Erica Schoenberger (1997) illuminates in her study of the multi-locational firm, TNCs encompass multiple agendas and seek out locations to serve those agendas. And, as our case studies demonstrate, firms are actively engaged in utilizing regional collective resources in their competitive endeavors. Whether that interaction works to the advantage of the region and it's residents is uncertain, even in the best of cases.

8 Remaking regions

Considering scale and combining investment and redistribution

In Chapters 6 and 7, we discussed prominent perspectives on the role of the region in the global economy – regional innovation systems and the learning region. While distinctive from one another in some important respects, both measure regional success with reference to the success of globally competitive firms that operate within the region. Their answer to "the regional question" – why do some regions attract investment while others do not – lies within the region. Regions that do not tailor their human and physical infrastructure to enable globally competitive firms to succeed will fail to attract investment and fall to the bottom of the regional pile. They will be identified as losers and dysfunctional. This is the contemporary account of uneven development – failure to "learn," failure to thrive.

This account is flawed in two important respects. First, it neglects how political policies and processes beyond the regional scale affect the ability of regional policy-makers to build the infrastructure to attract firms and to construct the capacity for economic development. Second, it equates the attraction of globally competitive firms and their networks with the development of a sustainable regional economy and the health and well-being of regional citizens. It ignores the potential gap between the demands of globally competitive TNCs and long-term regional economic development priorities. In earlier chapters of this book we described how that gap arises from the ability of TNCs to use regions strategically to simultaneously reduce costs and tap skilled labor pools and form a TNC agenda focused on sustainable competitive advantage.

In this chapter we explore the possibility for a different vision of regional economic development in the knowledge economy. This new vision requires "re-placing" the region in a political-economic context, understanding that the regional scale is being created, both politically and discursively. In this chapter as well as throughout the book our intention has been to stimulate thinking about the political processes and policies constructing the region, not only *in* the region but at all scales where political action is instrumental in constructing space. The processes remaking the region in the global economy include policies, such as deregulation and trade liberalization, that have unintended consequences for regional investment and economic development policy. The investment and disinvestment decisions set in motion by extra-regional political agendas have been deepened and accelerated by inter-regional competition. In the US, a highly fragmented governance system composed of units with significant independent authority, the states, and those dependent on the authority of the states, the cities, creates a basis for competition among places. The ability of firms to utilize that fragmentation, pitting cities, regions, and states against one another to increase their profits, derives from their enhanced political and economic power. The result is investment in some regions and disinvestment in others. It is this policy-enabled investment and disinvestment that underlies uneven development in the contemporary economy.

Second, we examine whether and how investment regionalism, in the forms associated with regional innovation systems and the learning region, can be combined with regional policies focused on distribution. The economic development goal of "distributive regionalism" is different from that of regional innovation system approaches. As developed in work such as that of Manuel Pastor, and Joan Fitzgerald and Nancey Green Leigh, distributive regionalism is predicated on the idea that regional success can only be measured in terms of the quality of life for all regional citizens, not only those employed in "innovative" global industries. In our formulation, the ability to effectively combine an investment orientation with good distributional outcomes requires a central focus on the workforce and on the region as a labor market.

Although some US regions are benefiting from their role as investment hubs for international firms aiming to cut costs and tap pools of flexible "brain power," that approach does not necessarily translate into broader measures of regional success or long-term economic development potential. Ultimately, it is impossible to achieve regional economic development without attention to the political processes that allocate resources and capacity at multiple geographic scales. And, healthy, sustainable regional economies require attention to both investment and the intra-regional allocation of resources. These two levels of policy action need to work in tandem.

Constructing the regional place – power, actors, and the discourse of competition

What is the region? As we laid out early in this book, what appears to be a simple and self-evident question is actually critical to interpreting the role of the region in the global economy, and how and why it is being re-made to serve particular purposes and interests. In fact, it is the absence of an accepted definition that makes the region malleable or flexible (not unlike its workforce). Vagueness serves a purpose – it leaves open questions of accountability, transparency, and citizenship.

As our case studies show, the processes constructing the regional landscape are not purely economic. They are also political. By extension, regional economic development policy is a question of political economy and an issue of scale.

Rather than a natural outcome of a now irrelevant nation-state, as Ohmae (1995) postulated, the emergence of the region as a privileged scale of activity is the result of power, politics, and policy. Key actors with political and economic interests are reshaping the scale at which we govern economic activities. This does not imply that the region as a place and a scale of action is not important but that we need to take a more critical stance toward its definition and to more carefully examine who is using it and for what purposes.

In other words, those interested in building economies with a capacity for sustainability as well as innovation, for investment and distributional equity, need to think beyond "globalization" as an explanation for how regions are integrated (or not integrated) in projects of capital accumulation. Policy decisions, such as those to privatize public transport or to deregulate communications distribution, will have consequences for access to and the quality and cost of services in different regions. Policies to enable industry concentration, through merger, acquisition, or hostile takeover will also affect firm location and investment decisions and, as our case study of the media industry demonstrates, affect the bargaining position of some firms vis-à-vis regions and their workforces.

Analysis attuned to concrete processes such as these can illuminate how the region's fortunes are shaped in a broader governance context, as economic actors attempt to position themselves in relationship to new opportunities and risks. It holds the potential to help us better understand how regional economic development and distributive equity are affected by the dynamic processes of inclusion and exclusion that emanate from beyond the boundaries of the region (Sheppard 2002). These perspectives build on the basic geographic idea that spaces, including regions, are not fixed but constantly being reorganized and reconstructed in conjunction with the dynamic social relations inherent to capitalist economies.

Actors matter. Just as we need to move toward definitions of "the region" that tell us about critical processes, we also need to identify the specific actors engaged in "making" regions. The process of constructing a new scale for the formal and informal negotiation of governance is not a neutral act. Participation in the "regional project" benefits some interests while it lessens the influence of others. For example, within the region, coalitions which draw their political support at the city or neighborhood scale – community-based organizations, local governments, and public agencies – see the region as a landscape that is determined by interests able to "jump scales," to use the multi-jurisdictional character of most US regions to evade democratic accountability.

At the vaguely defined regional scale, and in lieu of citizenship-based legitimacy and accountability, public–private partnerships have a powerful influence on the economic development landscape. "Quasi-governmental" policy is the hands of ad hoc groups, sometimes politically appointed and sometimes self-appointed, with an interest in either "the regional project" or in some specific aspect of it. We need to be clear about who is re-making the region and for what purposes.

As our case studies show, those with the most power over regional production and innovation systems as well as employment may eschew any interest in or attachment to the region. TNCs, such as Kodak in Rochester or GE-owned NBC Universal in Los Angeles adamantly resist any attachment to the region or "citizenship" responsibilities. They are, as they so often attest, "international firms." This does not inhibit them, however, from drawing on public tax subsidies for their infrastructure development projects or to increase their profits; taking advantage of specialized research or workforce development capacities in regional higher education institutions; or lobbying public officials to shape regional workforce conditions to meet their needs. While they may eschew a regional identity, we cannot ignore how their strategies affect the economic development potential and policy priorities in the region.

The role of firms as political actors is visible within the region but also is evident in the construction of inter-regional competition. In the US, a historically fragmented governmental system and devolution of responsibility for social welfare to states and localities create pressure to compete for "jobs" that could increase the tax base and pay for the ever-increasing burden of service provision. Within this context, an emphasis on regional winners and losers, and the importance of regional entrepreneurship has encouraged and accelerated subsidies to TNCs willing to locate activities in a region. As we noted in the introduction to this book, the significance of the region as a site for agglomeration economies may be out-weighed by its uses as a source of indirect and direct capital in the form of public subsidies.

Despite a broad awareness of the costs of inter-jurisdictional competition, subsidies to firms from sub-national state and local government have been steadily rising since the 1980s. A national policy initiative to curtail the inter-state and inter-regional competition was proposed by a team of national economic policy experts and published by the Minneapolis Federal Reserve Bank in 1996. They articulated both the problem of inter-jurisdictional competition and outlined solutions:

> Only Congress, under the Commerce Clause of the Constitution, has the power to enact legislation to prohibit the states from using subsidies and preferential taxes to compete with one another for businesses. Congress could enforce such a prohibition in a variety of ways. To name a few, it could tax real and imputed income from public subsidies, deny tax-exempt status to any public debt used to compete for businesses (there is already a limitation on the tax exempt status of certain kinds of state and local public debt) and impound federal funds payable to a state engaging in such competition.
>
> (Burstein & Rolnick 1996b)

With increasing corporate bargaining power, the problem of inter-jurisdictional competition in the US has worsened. Good Jobs First, an organization that tracks corporate subsidies, estimated that by 2004 Wal-Mart had received more than $1 billion in subsidies from state and local governments in the US In a detailed study, Good Jobs First found subsidies to Wal-Mart in the form of free or reduced-price land, infrastructure assistance, tax increment financing, property tax breaks, state corporate income tax credits, sales tax rebates, enterprise zone (and other zone) status, job training and worker recruitment fund, tax-exempt bond financing, and general grants (Mattera *et al.* 2004).

Although subsidies undermine market efficiency and are frequently fiscally irresponsible, corporations operating in the US continue to seek incentive packages and play regions against one other to get "a better deal." Transnational corporations in both old and new industries, including Boeing, Intel, IBM, Dell, Ford, and Honda, are adept players of the subsidy game and among the top recipients of subsidy deals, as seen in Table 8.1. As we described in our case study of the entertainment media industry, constructing inter-regional competition is not just a game reserved for manufacturing firms and retailers. Eli Lilly and Pfizer are also top recipients, as well as financial services firms like Wells Fargo, Vanguard Group, and Capital One (Mattera *et al.* 2004). After spending $500,000,000 on attracting an IBM plant or $100,000,000 for a Ford factory, a state will have limited resources for investment in infrastructure, schools, or a globally competitive regional innovation strategy focused on small firm networks. The subsidy game overwhelmingly favors transnational firm interests over those of small firms, who are in a weak position to attract subsidies because they lack political influence, and the lawyers and accountants required to craft the subsidy packages. Small firms, including those in innovative firm networks, pay the price for subsidies, however, in poor public services and higher taxes.

Examples of TNC use of public resources to reduce their risks and enhance their competitive position in the global economy represent cautionary tales about the role of power in and across regions. They suggest that policy-makers concerned with the long-term future of the region need to pursue a complex, multi-scalar agenda that includes working in coalitions to reduce inter-regional competition and advocating to ensure that the regional consequences of national policy are understood.

Recognizing the role that power can play in shaping regional agendas doesn't, however, obviate the necessity to find ways to respond to an economy in which there is more competition and in which knowledge-based production is central to economic development. One possibility is to combine the advantages of investment regionalism and agglomeration economies with a commitment to distributive regionalism with its emphasis on long-term sustainability and regional well-being. In the next sections we look at these regional projects and at how they could be joined.

Table 8.1 Economic development packages for firm location and expansion (1998–2004)

State	City	Company	Industry	Year	Amount
Alabama	Huntsville	Toyota	Motor vehicles	2001	$29,895,000
Alabama	Lincoln	Honda	Vehicles	1999	$158,000,000
Alabama	Lincoln	Honda (expansion)	Motor vehicles	2002	$89,700,000
Alabama	Montgomery	Hyundai	Motor vehicles	2002	$252,000,000
Arizona	Phoenix	USAA Insurance	Financial services	2000	$10,500,000
California	San Jose	Cisco Systems	Electronics	1998	$20,000,000
Florida	Palm Beach County	Scripps Research Institute	Biotechnology	2003	$310,000,000
Iowa	Des Moines	Wells Fargo	Financial services	2003	$45,000,000
Iowa	Indianapolis	Eli Lilly	Pharmaceuticals	1999	$214,000,000
Kentucky	Florence	Citicorp Credit Services	Financial services	2001	$26,700,000
Kentucky	Louisville	UPS	Package delivery	1998	$35,000,000
Michigan	Ann Arbor	Pfizer	Pharmaceuticals	2001	$84,200,000
Michigan	Flat Rock	Auto Alliance International* Ford/Mazda joint venture	Motor vehicles	2002	$133,000,000
Michigan	Lansing	General Motors	Motor vehicles	2000	$194,800,000
Michigan	Wayne	Ford Motor	Motor vehicles	2003	$106,800,000
Mississippi	Canton	Nissan	Motor vehicles	2000	$295,000,000
Mississippi	Canton	Nissan (expansion)	Motor vehicles	2002	$68,000,000
North Carolina	Winston-Salem	Dell	Computers	2004	$242,000,000
New York	Albany	International Sematech	Electronics	2002	$210,000,000
New York	East Fishkill	IBM	Electronics	2000	$503,750,000
New York	Fishkill	Gap	Retailing	1998	$20,000,000
New York	Fishkill & Albany	7 semi-conductor firms	Electronics	2004	$150,000,000
New York	New York	Pfizer	Pharmaceuticals	2003	$46,700,000
Pennsylvania	Malvern	Vanguard Group	Financial services	2000	$55,500,000
South Carolina	Charleston	Alenia/Vought	Aerospace	2004	$160,000,000
South Carolina	Greer	BMW	Motor vehicles	2002	$80,000,000
Tennessee	Nashville	Dell	Computers	1999	$166,000,000
Texas	Richardson	Texas Instruments	Electronics	2003	$135,000,000
Texas	Richardson	Countrywide Financial	Financial services	2004	$21,000,000
Texas	San Antonio	Toyota	Motor vehicles	2003	$400,000,000
Virginia	Dublin	Volvo	Motor vehicles	1999	$60,000,000
Virginia	Norfolk	Ford Motor	Motor vehicles	2001	$12,000,000
Virginia	Richmond	Capital One	Financial services	2000	$35,000,000
Virginia	Richmond	Philip Morris USA	Tobacco	2003	$28,000,000
Washington	Everett	Boeing	Aerospace	2003	$3,200,000,000

Source: Mattera *et al.*, 2004; *Good Jobs First* and *Site Selection Magazine*.

Investment regionalism

As we have already laid out in Chapters 6 and 7, investment regionalism emerged out of a literature in strategic management, centered on the work of Michael Porter, and a policy-oriented literature in economics and geography. Investment regionalism links continuous growth and the development of new technologies with regional competitiveness (Porter 2000) and emphasizes the importance of competitiveness over broader measures of regional economic health. Its claim to being "progressive" lies in its purported ability to enable a region to avoid deindustrialization and decline, and "a race to the bottom." By pursuing investments that enhance the capacity for continuous innovation, the region will be able to lessen or avoid the cost-based competition that is considered an inherent dimension of global market integration (Feldman *et al.* 2005; Porter 1990).

The centerpiece of investment regionalism, the regional innovation system, has captured the attention of policy-makers and politicians searching for ways to distinguish their regions as growth poles in the global economy. In some respects, the regional innovation system plays the role that large firms once played in local economies. Investments in institutional capacity, research and development, and workforce skills were once taken on by firms such as Kodak in Rochester, under the belief that the firm benefited from what became regional assets (Jacoby 1997). The firm continues to benefit from them but the cost of providing them has been largely transferred to the public sector, universities, and individual workers.

On the positive side, investment regionalism has opened opportunities for workforce intermediaries, including unions, to play a role in training and representing a new generation of skilled workers. As our Rochester case study shows, firms that could once set themselves apart from the regional labor market because of the depth of their internal labor market are finding themselves more dependent on how well the regional labor market and its training systems function. This opens the door to new bargaining and training arrangements that could give the workforce and their intermediaries more bargaining power.

Investment regionalism, through regional innovation systems, subsidizes research and development and provides infrastructure for firms with global markets in the expectation that the investment will lead to regional job growth. The proposition that regional innovation results in regional production, however, vastly over-simplifies economic processes (Dicken 2003). Recent studies of the computer, telecommunications, steel, and watch industries all tell complex stories about the reorganization of production processes and the role of large firms in shaping regional institutions and work (Angel & Engstrom 1995; Glasmeier 2000; Stone 1973, 2004; Wolf-Powers 2001b). The empirical evidence indicates that technological change produces a wide range of economic impacts – employment growth is only one scenario.

As a result, the question of whether innovation is connected to job growth is hotly contested. John Lovering argues, for example, that claims about investments in innovation leading to regional job growth should be met with skepticism:

> New regionalist accounts imply that new technologies and related science and technology or research and development activities ought to be the most important sources of employment growth. We might expect, therefore, that they would form the core of emergent labour markets. In reality, the opposite is the case: these sources of employment have been static or declining in most advanced industrial countries in recent years . . . It is a gross exaggeration to claim that innovation-related activities are the major source of new jobs, directly or indirectly, in regions or cities.
>
> (Lovering 1999)

Skepticism about a relationship between innovation and job growth has emerged for two important reasons: 1) the strong potential for a spatial disjuncture between innovation and production; and 2) the significance of process innovation, which reduces rather than increases jobs in knowledge-based industries.

As our analysis of the paradox of innovation in Chapter 6 shows, regions may have innovation capacity in their universities but lack the workforce skills or market access that could attract firms looking for production locations for the commercialized products. Even in cases where firms are established proximate to the innovation source, they may be forced to move closer to their investors if they accept venture capital or when they reach a critical size and need to tap a deeper labor pool. In the US, this is the time at which firms will move to the coastal megalopolitan regions.

Secondly, the presence of a regional innovation system, including a production component, may not produce large numbers of jobs. Small firms subcontracting to TNCs with global markets are under intense competitive pressure. They frequently cannot locate workers with the skills they require (as our study of Rochester indicated) because the skill base of the regional labor market has been eroded by worker exit and poor basic education. For these reasons, as well as cost competition, small specialized and innovative production firms turn to lean production methods in order to remain competitive. As a consequence, they create few jobs, even though the jobs they create may be good jobs (Christopherson *et al.* 2007).

This argues for an approach to investment regionalism focused on the needs and potential of small firms, including assistance to help them reach global markets on their own rather than through the conduit of the "lead firm," i.e. TNC.

Finally, there is the question of whether job growth, when it occurs, is associated with inequality. Some studies, for example Matt Drennan's recent work on the information economy, demonstrate a link between "high-tech" job growth and regional employment growth (Drennan 2002; Florida 2002c). While growth in knowledge-based industries may spur regional employment, much of the additional employment may occur in low-paying jobs in retail and service industries. As a consequence, critics, such as Lovering, see increasing income divergence, not employment growth, as the principle outcome of a focus on "high-tech" jobs (Lovering 2001; Martin & Sunley 1998). Empirical work reinforces this concern, suggesting that the presence of successful regional innovation capacity may not translate into a wider and deeper economic growth trajectory for the region unless steps are specifically taken to "spread the wealth" created by a regional innovation system (Christopherson *et al.* 2007).

As our analysis in Chapters 6 and 7 points out, investment regionalism in the form of regional innovation systems and learning regions is inadequate to the task of long-term regional development despite its considerable strengths in marshalling regional assets in the service of innovative capacity. That task requires more attention to the workforce and, by extension, to questions of distribution in the regional economy.

Distributive regionalism

Distributive regionalism emerged from a critical literature in political science focused on metropolitan and urban governance and from a long tradition in planning, which emphasized regional approaches to the challenges of urbanization. This tradition is exemplified by the regional plans for the New York City metropolitan region developed by the Regional Plan Association and in the work of Clarence Stein, Lewis Mumford, and other regional urbanists. In its contemporary manifestation, distributive regionalism is concerned with transparent and participatory governance, growth management in order to encourage efficiency through density, affordable housing, and tax base sharing to facilitate equitable access to critical services, such as schools and transportation, and to the amenities that urban regions can provide, such as parks and cultural activities (Dreier *et al.* 2001; Orfield 1997). It emphasizes the need to build institutions and place-based capacity. In distributive regionalism both research and practice are deeply engaged in questions of access, opportunity, and equity. At the heart of distributive regionalism are concerns about how planning practice and political decisions affect people, communities, and neighborhoods.

Distributive regionalism stands on a trajectory of planning theory and history engaged in questions of equity and distributional justice (Davidoff 1965; Friedmann & Weaver 1979; Markusen 1985, 1987). This "progressive regionalism" is closely connected to the concept "progressive cities" also focused on questions concerning distributional equity (Clavel 1986; Clavel & Wiewel 1991; Krumholz & Clavel 1994). Distributive regionalism, in fact, presents the possibility for a reemergence of "the progressive city" through a reconstitution of the boundaries (both conceptual and literal) of urban governance.

Distributive regionalism is also rooted in critiques of methodologically individualist approaches to urban governance, such as those represented by

Tiebout (1962) on the basis that they underrate the importance of equal access to certain services such as public transportation and good schools across a metropolitan region, and overestimate the ability of individuals to move freely from one urban neighborhood to another based on their preference for a certain package of services. Distributive regionalism, as its name implies, focuses on the consumption side of the economic ledger. Questions of how metropolitan areas develop economically or create jobs or build a tax base take second place to questions of equity and social justice. To the extent that questions of economic development are addressed, they begin from a presumption that more equitable and transparent government will ultimately produce a more economically competitive and attractive region.

When distributive regionalists focus on the region, it is on the city-region. A decade of sophisticated research has demonstrated that the city's claim on city-region resources is slipping as the political and economic power of the suburbs grows and the ability of urban actors to leverage a "fair share" of that growth is diminished through lax regulatory and legal enforcement of existing protections and a generally antiquated and fragmented urban governance regime. Political scientists and public policy professionals, in the tradition of urban politics, have analyzed the barriers to regional equity and located it in fragmented governance and a growth coalition of politicians, developers, and unions that benefit individually and collectively from expansion of the metropolis on green field sites (Dreier *et al.* 2001; Swanstrom 2001; Weir *et al.* 2005).

Simultaneously, urban and regional planners have committed to their own regional project, documenting the problems of the city-region – social, environmental, economic, spatial – and the ways in which the failure to adopt multi-jurisdictional governance approaches within the region exacerbates the problems facing individual regions and the metropolitan region as a whole.

For example, the lack of affordable housing magnifies the effects of discrimination and poverty in urban neighborhoods but also increases the likelihood that job seekers will have to travel long distances to find jobs (Goetz 2000; Pendall 2000). The failure of regional policy to address issues of access and transportation contributes to a city form in which household location is determined by a combination of socio-economic status and racial and ethnic identity that dictates opportunity (Grengs 2002; Immergluck 1998). While transportation services are perhaps the most regionalized, due to federal requirements for metropolitan transportation planning, they are flawed in their implementation as issues of equity and distribution take a back seat to questions of efficiency and engineering and opportunities for real estate development (Vogel 2002).

Environmental justice has emerged as another research area supporting the move towards a progressive "distributive" regionalism. The concentration of environmental disamenities, like bus depots, sewage treatment plants, and power generation and transmission facilities in densely populated poor and minority neighborhoods within the city limits adds health and safety concerns to the already uneven distribution of access and opportunities to regional amenities (Pulido 2000; Rast 2006).

Connecting investment and distributive regionalism through a focus on the workforce

Because regions are defined variously for political and economic purposes, multiple and overlapping definitions may exist at the same time. Perhaps ironically, the most obvious functional definition, that of the regional labor market, is rarely utilized as a definition around which to build public policy. Regional labor markets do not respect administrative boundaries. They cross city, county, and state lines. They intrinsically require cooperation and innovative policy thinking about labor and about the institutions that contribute to a skilled adaptive workforce. That is, perhaps, why they are ignored in favor of administrative regions, which, in the US, are amalgams of jurisdictions, designed for different political purposes. They are typically overlapping, different configurations of a set of jurisdictions, designated as regions for the purposes of educational programs, economic development, and environmental planning. In the US, regions are defined to capture the political allegiance of local officials or to create new political bases. They are also designed to exclude or undermine the power of areas, such as central cities. This exclusion enables regional political actors to avoid confronting the consequences of spatial or social inequality and to maintain a "coherent" power base.

A labor market definition of the region raises uncomfortable questions about the labor force as people rather than production inputs. Who are they? Where do they work? How far do they travel to work? Where do they live? How does their work and residence pattern intersect with education and training institutions? This kind of questioning moves us away from less challenging questions about the workforce, such as those embodied in the concept of "the creative class," to more challenging questions about the jobs held by the majority of people. Ideally a labor market orientation would also raise questions about how the costs of employment are accounted for. Who pays for skill acquisition? Who pays for commuting costs? Who pays for health insurance?

A focus on regional labor markets has the potential to create a link between the questions raised in the distributive regionalist "project," and the demand-oriented priorities of investment regionalism (Markusen 2004; Peck 1996). This potential is represented in a body of research that examines the complex intersection between economic development and community capacities (Chapple 2006; Wolf-Powers 2001a). In addition, this research, which straddles investment regionalism and distributive regionalism, addresses a key criticism of regional economic development organized around innovation systems – the creation of jobs beyond those in the firm network.

A labor market and workforce-oriented regionalism leads to questions concerning the broad range of people in a labor market, their skills and long-term potential for sustaining a livelihood (Giloth 1998; Harrison 1972). It also expands the range of institutions that are deemed necessary to support long-term economic development to the intermediaries that connect people to jobs and careers (Benner 2003). Finally, it recognizes that skills – whether in services or manufacturing – are essential for career ladders, allowing people to get into work, build productivity, and move up through experience (Fitzgerald & Carlson 2000). Empirical studies demonstrate that skilled labor (including high-skilled, medium-skilled, and generally skilled workers) are at the center of competitive industry strategies and essential to sustainable regional growth (Christopherson & Clark 2006; Rutherford & Holmes 2006).

The search for ways to connect investment regionalism, centered on regional innovation systems, with distributive regionalism, centered on equity, access, and quality of life is a search for a model of sustainable economic development.

Conclusions: an alternative future for regional development

At the heart of the issues raised in this chapter and in this book is a normative question about the goals of economic development. Joan Fitzgerald and Nancey Green Leigh have argued that economic development should increase standards of living, reduce inequality, and promote sustainable resource use and production (Fitzgerald & Leigh 2002). Under current conditions, neither the policy models of distributive nor investment regionalism meet the challenge of mitigating inequalities while growing a sustainable regional economy in a competitive world. If this is the goal of progressive regionalism, then regionalists need to develop a theoretical framework and policy strategies that both articulate the problems more clearly and propose compelling solutions.

A successful progressive regionalism must account for investment and growth. Regional innovation systems can, at least potentially, tie investment to the region rather than transferring it to the production networks of transnational firms through subsidies or tax breaks. Regional innovation systems can also provide political support for public investment in the quality of life and learning institutions that provide the basis for greater access and opportunity.

What regional innovation systems do not do is provide a vision for mitigating uneven development, either within or between regions. They also do not propose mechanisms for developing the skills and competencies of the entire labor force, *in situ*, and over time (Peck 1992). And finally, regional innovation systems raise questions about legitimacy and governance in "regional" jurisdictions.

Regions need innovative and creative firms to compete in a global economy. Traditional economic development valued firms for their ability to create jobs. In the contemporary economy, incentives have shifted to foster increases in productivity and wealth in terms of exports. Ultimately, however, economic development has to balance innovation and job creation. Investments in regional institutions and quality of life contribute to both agendas.

Regions are places, not simply sites of production. Regional economic development discourse too often conflates regions and firms, without distinguishing between the two. The fate of firms is perceived as inexorably tied to the region and vice versa. However, TNCs have maintained and expanded their ability to shift production among regions and to alter what they produce and how they produce it. The partnership of the TNC and the region is unequal. The TNC's interests take priority over those of the region. The power relationship – between production and place – remains at the heart of the region's competitive dilemma just as it was for the city. Only the scale has changed.

Notes

Chapter 1

1 This is not a wholesale rejection of the role and importance of a regional social infrastructure. The idea that informal rules and habits aid coordination of economic actors under conditions of uncertainty is borne out by numerous empirical studies. Also unproblematic is the idea that this social infrastructure and its informal rules are differentiated by regions and function in regionally specific ways. Again, comparisons among regional industries, such as those in the media, demonstrate considerable difference in industrial cultures across regions (Batt *et al.* 2001).

Chapter 2

1 The varieties-of-capitalism approach emerged with work by Michel Albert (1991) and has mushroomed into a significant comparative literature examining how so-called "global" processes are manifested differently in different nations because of differences in their social, political, and economic institutions.
2 From the corporation perspective, this is portrayed as representing their interests in a pluralistic democratic political system.
3 In some national economies there are measures to enforce accountability for claims on the public purse or to weigh and temper the costs of transnational firm agendas. In the U.S., however, firm influence on the political-economy is not tempered and so we see potential for firm influence in a clearer light. This being said, firms do attempt to rationalize their claims on the public purse and influence on regulation.
4 This perspective stands in strong contrast to prevailing conceptions of governance in the management literature, which, until recently, has equated governance with government. Government is in a completely separate realm from private sector actors and, in general, is portrayed as constraining and limiting firm action through, for example, regulation. The governmental realm is, however, also depicted as pluralist, that is affected equally by multiple interests. Firms, constituting one of those interests, vie along with other interests to influence the process through which aggregate preferences are determined.

Chapter 4

1 The headquarters of Xerox moved to Connecticut but a research and production presence remains in Rochester.
2 The primary interviewer for the survey project was Wyeth Friday. The interviews with the private labor market intermediaries were conducted by David Perkins.

3 The identification of firms, including finding contact information, was based on the lists of industry associations, public agencies, and publicly available data from the firms themselves.

4 See Osterman 1999: 68–69. Establishment survey cited in Osterman has the same response rate as the photonics survey.

5 See (U.S. Bureau of Labor Statistics 2003): "In 2002, 13.2 percent of wage and salary workers were union members, down from 13.4 percent (as revised) in 2001, the U.S. Department of Labor's Bureau of Labor Statistics reported today. The number of persons belonging to a union fell by 280,000 over the year to 16.1 million in 2002. The union membership rate has steadily declined from a high of 20.1 percent in 1983, the first year for which comparable union data are available . . . Four states had union membership rates over 20.0 percent in 2002 – New York (25.3 percent), Hawaii (24.4 percent), Alaska (24.3 percent), and Michigan (21.1 percent). This is the same rank order as in 2001. All four states have had rates above 20.0 percent every year since data became regularly available in 1995."

6 This chapter draws on a series of interviews with managers and owners of private, for-profit labor market intermediaries in Rochester. The interviews focused on how LMIs interacted with the optics, imaging, and photonics industry (Clark 2004).

7 Several branch temporary agencies are run by former Kelly girls as is the Industrial Management Council's labor supply agency.

8 The technology heavy NASDAQ stock index fell from over 5,000 to below 1,500 and the telecommunications went into a serious series of bankruptcies.

9 Interview with Industrial Management Council, Spring 2002.

10 U.S. Bureau of Labor Statistics Data (1997–2003). For the chart the classifications used included SIC 736 and NAICS 5613.

11 Interview with an Adecco branch in Rochester, Spring 2002.

12 Interview with an Adecco branch in Rochester, interview Spring 2002.

13 See Phelps 1997: 27: "It is a fallacy that normal economic processes operate to pull up wage rates at the low end relative to those in the middle – that is, to erode inequality."

14 Interview with with the Industrial Management Council, Spring 2002.

15 Interview with Manpower, Inc. in Rochester, Spring 2002.

16 Interview with Burns Personnel, Rochester, Spring 2002.

17 Burns Personnel described reverse bidding as a process in which Xerox or Kodak electronically posts a labor contract. Suppliers then bid on-line a bill rate for that contract, trying to manage a bid lower than their competitors but still profitable based on the volume of the contract.

18 Interview with Gemini Personnel, Spring 2002.

19 Interview with Gemini Personnel, Spring 2002.

20 Interview with Gemini Personnel, Spring 2002.

21 Interview with Industrial Management Council, Spring 2002.

22 Interview with Manpower Inc., Spring 2000.

23 Interview with Gemini Personnel, Spring 2002.

Chapter 5

1 Television production appears to be outstripping feature film production as a source of employment in the media entertainment industries in the United States (California State Department of Employment, 2005; Christopherson 2005).

Chapter 7

1 Any characterization of an economy as complex as that of that of the U.S. is always something of a caricature because there are firms that don't operate according to "the rules." Our objective here is to emphasize the importance of the incentive structure rather than every outcome.

2 Although U.S. universities would like to cooperate in order to, for example, reduce bidding for top students, that opportunity has been precluded by anti-trust law. U.S. universities are, in essence, forced to compete with one another, thus driving up their costs.

Bibliography

Acs, Z. J., Audretsch, D. B., & Feldman, M. P. (1994) R & D spillovers and recipient firm size. The Review of Economics and Statistics, 76(2), 336.

Albert, M. (1991) Capitalisme contre capitalisme. Paris: Le Seuil.

Allen, J. (2003) Lost geographies of power. Oxford: Blackwell Publishing.

Allen, J. (2004) The whereabouts of power: Politics, government, and space. Geografiska Annaler, 86b(1), 17–30.

Amin, A. (1994) Post-Fordism: A reader. Oxford; Cambridge, Mass.: Blackwell.

Amin, A. (1999) An institutionalist perspective on regional economic development. International Journal of Urban and Regional Research, 23(2), 365.

Amin, A., & Thrift, N. (1992) Neo-marshallian nodes in global networks. International Journal of Urban and Regional Research, 16(4), 571.

Angel, D. P. (2002) Inter-firm collaboration and technology development partnerships within US manufacturing industries. Regional Studies, 36(4), 333.

Angel, D. P., & Engstrom, J. (1995) Manufacturing systems and technological change: The U.S. personal computer industry. Economic Geography, 71(1), 79.

Aoyama, Y., & Schwarz, G. (2006) The myth of Wal-Martization: Retail globalization and local competition in Japan and Germany. In S. D. Brunn (ed.), WalMart World, The world's biggest corporation in the global economy. London: Routledge.

Asheim, B. (1992) Flexible specialisation, industrial districts and small firms: A critical appraisal. In H. Ernste & V. Meier (eds), Regional development and contemporary industrial response – extending flexible specialization (pp. 45–64) London: Belhaven Press.

Asheim, B. T., & Isaksen, A. (2002) Regional innovation systems: The integration of local 'sticky' and global 'ubiquitous' knowledge. Journal of Technology Transfer, 27(1), 77.

Audretsch, D. (2004) Sustaining innovation and growth: Public policy support for entrepreneurship. Industry and Innovation, 11(3), 167.

Audretsch, D., & Feldman, M. (2003) Small-firm strategic research partnerships: The case of biotechnology. Technology Analysis & Strategic Management, 15(2), 273.

Badaracco, J. (1991) The boundaries of the firm. In A. Etzioni & P. Lawrence (eds), Socio-economics: Toward a new synthesis. New York: Armonk.

Bagdikian, B. H. (2000) The media monopoly (6th edn) Boston: Beacon Press.

Bania, N., Randall, W., Eberts, M., & Fogarty (1993) Universities and the startup of new companies: Can we generalize from Route 128 and Silicon Valley? The Review of Economics and Statistics, 75(4), 761–766.

Barnes, T. J. (2004) The rise (and decline) of American regional science: Lessons for the new economic geography? Journal of Economic Geography, 4, 107–129.

Barnes, T. J., & Gertler, M. S. (1999) The new industrial geography: Regions, regulations and institutions. London; New York: Routledge.

Barnes, T., Hayter, R., & Grass, E. (1990) MacMillan Bloedel: Corporate restructuring and employment change. In M. de. Smidt & E. Wever (eds), The corporate firm in a changing world economy. London: Routledge.

Bartik, T. (1991) Who benefits from state and local economic development policies? Kalamazoo, Michigan: The Upjohn Institute for Employment Research.

Batt, R. S., Christopherson, S., Rightor, N., & Van Jaarsveld, D. (2001) Net working, work patterns and workforce policies for the new media industry. Washington D.C.: Economic Policy Institute.

Befort, S. (2003) Revisiting the black hole of workplace regulation: A historical and comparative perspective of contingent work. Berkeley Journal of Employment and Labour Law, 24(1).

Belzowski, B., Flynn, M., Richardson, B., Sims, M., & VanAssche, M. (2003) Harnessing knowledge: The next challenge to inter-firm cooperation in the North American auto industry. Ann Arbor: University of Michigan Transportation Research Institute.

Beneria, L., & Santiago, L. E. (2001) The impact of industrial relocation on displaced workers: A case study of Cortland, New York. Economic Development Quarterly, 15(1), 78.

Benner, C. (2001) Global trends in flexible labor. Regional Studies, 35(9), 881.

Benner, C. (2003) Labor flexibility and regional development: The role of labour market intermediaries. Regional Studies, 37(6), 621.

Benneworth, P. (2006) Creating economic possibilities in old industrial regions: The role of the university. Newcastle Upon Tyne: Centre for Urban and Regional Development Studies, University of Newcastle Upon Tyne, England.

Berger, S. (2006) How we compete: What companies around the world are doing to make it in today's global economy (1st edn). New York: Currency Doubleday.

Berk, G. (1994) Alternative tracks: The constitution of American Industrial Order, 1865–1917. Baltimore, Md.: Johns Hopkins University Press.

Berk, G., & Swanstrom, T. (1995) The power of place: Capital (im)mobility, pluralism, and regime theory. Paper presented at the Annual meeting of American Political Science Association, Chicago, Ill.

Bianchi, P. (1994) Technology and human resources in Europe after Maastricht. International Journal of Technology Management, 9(3,4), 314.

Blackwell, R. (2003, October 9) Canadian film industry worries. Toronto Globe and Mail.

Bluestone, B., & Harrison, B. (1982) The deindustrialization of America: Plant closings, community abandonment, and the dismantling of basic industry. New York: Basic Books.

Boekema, F., Morgan, K., Bakkers, S., & Rutt, R. (eds) (2000) Knowledge, innovation and economic growth: The theory and practice of learning regions. Cheltenham: Edward Elgar.

Boschma, R., & Lambooy, J. (2002) Knowledge, market structure, and economic coordination: Dynamics of industrial districts. Growth and Change, 33(3), 291.

Bozeman, B. (2000) Technology transfer and public policy: A review of research and theory. Research Policy, 29(4,5), 627.

Bozeman, B., & Boardman, C. (2004) The NSF engineering research centers and the university-industry research revolution: A brief history featuring an interview with Erich Bloch. Journal of Technology Transfer, 29(3–4), 365.

Braczyk, H., Cooke, P., & Heidenreich, M. (1998) Regional innovation systems: The role of governances in a globalized world. London; Bristol, Pa.: UCL Press.

Brenner, N. (2004) New state spaces, urban governance and the rescaling of statehood. Oxford: Oxford University Press.

Brenner, N. (2005) New state spaces, urban governance and the rescaling of statehood. Oxford: Oxford University Press.

Brunn, S. (2006) Wal-Mart world: The world's biggest corporation in the global economy. New York: Routledge.

Burstein, M., & Rolnick, A. (1996a) Congress should end the economic war for sports and other business. Region 10, No. 2, 35–36.

Burstein, M., & Rolnick, A. (1996b) The economic war among the States: Federal Reserve Bank of Minneapolis.

California State Department of Employment (2005) Biennial report on the motion picture and television industry in California. Sacramento, Calif.: California State Department of Employment Development.

Canadian Film and Television Production Association (2003) Producers applaud Feds as they boost tax credit. Paper presented at the Press Release. from www.cftpa.ca.

Canadian Film and Television Production Association. (2005) Submission to the standing committee on finance pre-budget consultation.

Cappelli, P. (1997) Change at work. New York: Oxford University Press.

Cappelli, P., & Neumark, D. (2001) External job churning and internal job flexibility (Working Paper 8111) Cambridge, Mass.: National Bureau of Economic Research.

Carnoy, M., Castells, M., & Benner, C. (1997) Labour markets and employment practices in the age of flexibility: A case study of Silicon Valley. International Labour Review, 136(1), 27.

Chapple, K. (2006) Overcoming mismatch: Beyond dispersal, mobility, and development strategies. Journal of the American Planning Association, 72(3).

Christopherson, S. (1989) Flexibility in the U.S. service economy and the emerging spatial division of labour. Transactions of the Institute of British Geographers, 14, 131–143.

Christopherson, S. (1993) Market rules and territorial outcomes: The case of the United States. International Journal of Urban and Regional Research, 17(2), 274.

Christopherson, S. (1996) Industrial relations in an international industry, film production. In L. Gray & R. Seeber (eds), Under the stars, industrial relations in the entertainment media industries. Ithaca, N.Y.: Cornell University Press.

Christopherson, S. (1999) Review of Richard Sennett, The corrosion of character: The personal consequences of work in the new capitalism. International Journal of Urban and Regional Research, 23(4), 802.

Christopherson, S. (2002a) Project work in context: Regulatory change and the new geography of the media. Environment and Planning A 34, 2003–2015.

Christopherson, S. (2002b) Why do national labor market practices continue to diverge in the global economy? The "missing link" of investment rules. Economic Geography, 78(1), 1.

Christopherson, S. (2004) Review of Jefferson Cowie and Joseph Heathcott (eds.) Beyond the Ruins, The Meaning of Deindustrialization. Journal of the American Association of Planning, 70(4 (Autumn)), 487–488.

Christopherson, S. (2005) Building the future of film, television, and commercial production in New York. Unpublished work, Cornell University.

Christopherson, S., & Clark, J. (2000) Working Paper: Restructuring a specialized economy: The Rochester regional labor market.Unpublished manuscript, Ithaca, NY.

Christopherson, S., & Clark, J. (2007a) The politics of firm networks: How large firm power limits small firm innovation. Geoforum, 38(1), 1–3.

Christopherson, S., & Clark, J. (2007b) Power in firm networks: What it means for regional innovation systems (forthcoming Regional Studies)

Christopherson, S., & Storper, M. (1985) The changing organization and location of the motion picture industry. Graduate School of Architecture and Urban Planning, University of California, Los Angeles.

Christopherson, S., & Storper, M. (1986) The city as studio; the world as backlot: The impact of vertical disintegration on the location of the motion picture industry. Environment and Planning D: Society and Space, 4, 305–320.

Christopherson, S., & Storper, M. (1989) The effects of flexible specialization on industrial politi. Industrial & Labor Relations Review, 42(3), 331.

Christopherson, S., Brown, W., & Rightor, N. (2007) Advanced manufacturing in New York's Southern Tier: A report to the New York State Association of Counties. Albany, N.Y.: New York State Association of Counties.

Christopherson, S., Gray, L. S., Figueroa, M., Parrott, J., Richardson, D., & Rightor, N. (2006) New York's big picture: Assessing New York's position in film, television and commercial production. New York: The New York Film, Television and Commercial Initiative.

Clark, G. L. (1989) Unions and communities under siege: American communities and the crisis of organized labor. Cambridge, England; New York: Cambridge University Press.

Clark, G. L., & Wrigley, N. (1997) Exit, the firm and sunk costs: Reconceptualizing the corporate geography of disinvestment and plant closure. Progress in Human Geography, 21(3), pp 338–358.

Clark, J. (2004) Restructuring the region: The evolution of the optics and imaging industry in Rochester, New York. Unpublished Ph.D., Cornell University, Ithaca, N.Y.

Clavel, P. (1986) The progressive city: Planning and participation, 1969–1984. New Brunswick, N.J.: Rutgers University Press.

Clavel, P., & Wiewel, W. (1991) Harold Washington and the neighborhoods: Progressive city government in Chicago, 1983–1987. New Brunswick, N.J.: Rutgers University Press.

Coe, N. M. (2001) A hybrid agglomeration? The development of a satellite-marshallian industrial district in Vancouver's film industry. Urban Studies, 38, 1,753–1775.

Cooke, P. (2002) Regional innovation systems and regional competitiveness. In M. Gertler & D. Wolfe (eds), Innovation and social learning. New York: Palgrave Macmillan.

Cooke, P. (2004) Regional innovation systems: An evolutionary approach. In P. Cooke (ed.), Regional innovation systems (2nd edn). London: Routledge.

Cooke, P. (2005) Regional transformation and regional disequilibrium: New knowledge economies and their discontents. In Fuchs & P. Shapira (eds), Rethinking regional innovation and change. Berlin: Springer.

Cooke, P., & Morgan, K. (1998) The associational economy: Firms, regions, and innovation. Oxford England; New York: Oxford University Press.

Cooke, P. & Piccaluga, A. (eds) (2004) Regional economies as knowledge laboratories. Cheltenham: Edward Elgar.

Council on Competitiveness (2007) Regional innovation: rationale. http://www.compete.org/nri/

Coursey, D., & Bozeman, B. (1992) Technology transfer in U.S. government and university laboratories: Advantages and disadvantages for participating laboratories. IEEE Transactions on Engineering Management, 39(4), 347.

Cowie, J. R., & Heathcott, J. (2003) Beyond the ruins: The meanings of deindustrialization. Ithaca, N.Y.: ILR Press.

Crevoisier, O. (1999) Two ways to look at learning regions in the context of globalization: The homogenizing and particularizing approaches. GeoJournal, 49(4), 353.

Crevoisier, O. (2004) The innovative milieus approach: Toward a territorialized understanding of the economy? Economic Geography, 80(4), 367.

Danielson, D. (2005) How corporations govern: Taking corporate power seriously in transnational regulation and governance. Harvard International Law Journal, 46(2), 411–425.

Davidoff, P. (1965) Advocacy and pluralism in planning. Journal of the American Institute of Planners, 31(4), 331–338.

DeFilippis, J. (2001) The myth of social capital in community development. Housing Policy Debate, Vol. 12(4), 781–806.

Dicken, P. (1988) Global shift: Industrial change in a turbulent world. London: Paul Chapman Publishing.

Dicken, P. (1998) Global shift: Transforming the world economy. New York: Guilford Press.

Dicken, P. (2003) Global shift: Reshaping the global economic map in the 21st century (4th edn). New York: Guilford Press.

Dicken, P., & Malmberg, A. (2001) Firms in territories: A relational perspective. Economic Geography, 77(4), 345.

Dicken, P., Kelly, P., Olds, K. & Yeung, H. (2001) Chains and networks, territories and scales: Towards a relational framework for analysing the global economy. Global Networks, 1(2), 89–112.

Dobbin, F., & Dowd, T. (1997) How policy shapes competition: Early railroad foundings in Massachusetts. Administrative Science Quarterly, 42, 501–529.

Doeringer, P. B., & Piore, M. J. (1971) Internal labor markets and manpower analysis. Lexington, Mass.: Heath.

Doremus, P., Keller, W., Pauly, L., & Reich, S. (1998) The myth of the global corporation. Princeton, N.J.: Princeton University Press.

Dreier, P., Mollenkopf, J., & Swanstrom, T. (2001) Place matters: Metropolitics for the 21st century. Lawrence, Kansas: University Press of Kansas.

Drennan, M. P. (1998) Economic change in Western New York 1989–1998. Buffalo, N.Y.: Buffalo Branch, Federal Reserve Bank of New York.

Drennan, M. P. (2002) The information economy and American cities. Baltimore: Johns Hopkins University Press.

Dunford, M. (2003) Theorizing regional economic performance and the changing territorial division of labour. Regional Studies, 37(8), 839.

Edelman, L. (2004) Rivers of law and contested terrain: A law and society approach to economic rationality. Law & Society Review, 38(2).

Egeln, J., Gottschalk, S., & Rammer, C. (2004) Location decisions of spin-offs from public research institutions. Industry and Innovation, 11(3), 207.

Ehrenberg, R., Rizzo, M., & Jakubson, G. (2003) Who bears the cost of science at universities (No. WP35) Ithaca, N.Y.: Cornell Higher Education Research Institute Working Paper.

Eisenmann, T. R., & Bower, J. L. (2000) The entrepreneurial M-form: Strategic integration in global media firms Organization Science, 11(3), 348.

Elmer, G., & Gasher, M. (2005) Contracting out Hollywood, runaway productions and foreign location shooting. Toronto: Rowman & Littlefield.

Entertainment Economy Institute, & PMR Group (2005) California's entertainment workforce: Employment and earnings (1991–2002). Los Angeles: The Entertainment Economy Institute, PMR Group.

Entertainment Industry Development Corporation (2001) MOWs – A three year study. Los Angeles: Entertainment Industry Development Corporation.

Epstein, E. J. (2005) The big picture, the new logic of money and power in Hollywood. New York: Random House.

Erbil, T. (2006) The effects of privatization on the development of national innovation systems in developing countries: A case study of the Turkish telecommunications industry. Unpublished Dissertation, Cornell University, Ithaca, N.Y.

Ernst, H., Witt, P., & Brachtendorf, G. (2005) Corporate venture capital as a strategy for external innovation: An exploratory empirical study. R & D Management, 35(3), 233–242.

Feldman, M. P. (2000) Location and innovation: The new economic geography of innovation, spillovers, and agglomeration. In G. L. Clark, M. P. Feldman & M. Gertler (eds), The Oxford handbook of economic geography. New York: The Oxford University Press.

Feldman, M. P. (2001) The entrepreneurial event revisited: Firm formation in a regional context. Industrial and Corporate Change, 10(4), 861.

Feldman, M. P., & Desrochers, P. (2003) Research universities and local economic development: Lessons from the history of the Johns Hopkins University. Industry and Innovation, 10(1), 5.

Feldman, M. P., Francis, J., & Bercovitz, J. (2005) Creating a cluster while building a firm: Entrepreneurs and the formation of industrial clusters. Regional Studies, 39(1), 129.

Feldman, M. P., Link, A. N., & Association for Public Policy Analysis and Management Conference. (2001) Innovation policy in the knowledge-based economy (Vol. 23) Boston: Kluwer Academic Publishers.

Feller, I. (1999) The industrial perspective. New York: New York Academy of Sciences.

Film and Television Action Committee (2004) We are creating the jobs your children want. Los Angeles: Film and Television Action Committee.

Fitzgerald, J., & Carlson, V. (2000) Ladders to a better life. The American Prospect, 11(15), 54.

Fitzgerald, J., & Leigh, N. G. (2002) Economic revitalization: Cases and strategies for city and suburb. Thousand Oaks, California: Sage Publications.

Florida, R. (2002a) The economic geography of talent. Annals of the Association of American Geographers, 92(4), 743–755.

Florida, R. (2002b) The learning region. In M. Gertler & D. Wolfe (eds), Innovation and social learning: Institutional adaptation in an era of technological change. New York: Palgrave Macmillan.

Florida, R. (2002c) The rise of the creative class: And how it's transforming work, leisure, community, and everyday life. New York: Basic Books.

Florida, R. (2005) The flight of the creative class: The new global competition for talent. New York: Harper Collins.

Florida, R., & Kenney, M. (1990) Silicon Valley and Route 128 won't save us. California Management Review, 33, 68–88.

Friedman, T. L. (2005) The world is flat. New York: Farrar, Strauss and Giroux.

Friedman, T. L. (2006) The world is flat: A brief history of the twenty-first century (1st updated and expanded edn) New York: Farrar, Straus and Giroux.

Friedmann, J., & Weaver, C. (1979) Territory and function: The evolution of regional planning. London: E. Arnold.

Fujita, M., Krugman, P. R., & Venables, A. (1999) The spatial economy: Cities, regions and international trade. Cambridge, Mass.: MIT Press.

Galbraith, J. K. (1967) The new industrial state. Boston: Houghton Mifflin.

Galbraith, J. K. (1977) The age of uncertainty. Boston: Houghton Mifflin.

Galbraith, M. W. (2004) Adult learning methods: A guide for effective instruction (3rd edn) Malabar, Fl.: Krieger Pub. Co.

Gereffi, G. (1996) Global commodity chains: New forms of coordination and control among nations and firms in international industries. Competition and Change, 4(1), 427–439.

Gereffi, G., Humphrey, J., & Sturgeon, T. (2005) The governance of global value chains. Review of International Political Economy, 12(1), 78–104.

Gertler, M. S. (2003) Tacit knowledge and the economic geography of context, or the undefinable tacitness of being (there) Journal of Economic Geography, 3(1), 75.

Gertler, M. S. & Levitte, Y. M. (2005) Local nodes in global networks: the geography of knowledge flows in biotechnology innovation. Industry and Innovation, 12(4): 487–507.

Gertler, M. S., & Wolfe, D. A. (2002) Innovation and social learning: Institutional adaptation in an era of technological change. New York: Palgrave Macmillan.

Gierzynski, A. (2000) Money rules: Financing elections in America. Boulder, Colo.: Westview Press.

Giloth, R. (1998) Jobs & economic development: Strategies and practice. Thousand Oaks, Calif.: Sage Publications.

Gilson, R. J. (1999) The legal infrastructure of high technology industrial districts: Silicon Valley, Route 128, and covenants not to compete. New York University Law Review, 74(3), 575.

Glasmeier, A. (1991) Technological discontinuities and flexible production networks: The case of Switzerland and the world watch industry. Research Policy, 20(5), 469.

Glasmeier, A. (2000) Manufacturing time: Global competition in the watch industry, 1795–2000. New York: Guilford Press.

Glasmeier, A. (2002) Knowledge, innovation, and economic growth. Association of

American Geographers. Annals of the Association of American Geographers, 92(3), 588.

Glasmeier, A. (2004) Geographic intersections of regional science: Reflections on Walter Isard's contributions to geography. Journal of Geographical Systems, 6(1), 27–41.

Goetz, E. G. (2000) Fair share or status quo?: The Twin Cities Livable Communities Act. Journal of Planning Education and Research, 20(1), 37–51.

Goodman, R. (1979) The last entrepreneurs: America's regional wars for jobs and dollars. New York: Simon and Schuster.

Gourevitch, P. A., & Shinn, J. (2005) Political power and corporate control: The new global politics of corporate governance. Princeton: Princeton University Press.

Grabher, G. (1993) The embedded firm: On the socioeconomics of industrial networks. London; New York: Routledge.

Graham, S. (1998) The end of geography or the explosion of place?: Conceptualizing space, place and information technology. Progress in Human Geography 22: 165–185.

Grantham, G., & MacKinnon, M. (eds) (1994) Labour market evolution: The economic history of market integration, wage flexibility, and the employment relationship. London: Routledge.

Gray, L., & Seeber, R. (1996) Under the stars, essays on labor relations in arts and entertainment. Ithaca, N.Y: Cornell University Press.

Gray, M., Golob, E., & Markusen, A. (1996) Big firms, long arms, wide shoulders: The 'Hub-and-Spoke' industrial district in the Seattle region. Regional Studies, 30(7), 651.

Greider, W. (2003) The soul of capitalism: Opening paths to a moral economy. New York: Simon & Schuster.

Grengs, J. (2002) Community-based planning as a source of political change: The transit equity movement of Los Angeles' bus riders union. American Planning Association. Journal of the American Planning Association, 68(2), 165.

Grimshaw, D., & Rubery, J. (2005) Inter-capital relations and the network organisation: Redefining the work and employment nexus. Cambridge Journal of Economics, 29, 1027–1051.

Hagenbauch, B. (2006) U.S. manufacturers desperate for skilled workers. USA Today December 5, 1.

Hall, P. A., & Soskice, D. W. (2001) Varieties of capitalism: The institutional foundations of comparative advantage. Oxford; New York: Oxford University Press.

Hansen, S. B., Ban, C., & Huggins, L. (2003) Explaining the "brain drain" from older industrial cities: The Pittsburgh region. Economic Development Quarterly, 17(2), 132.

Hanson, S. (2003) The entrepreneurship dynamic: Origins of entrepreneurship and the evolution of industries. Economic Geography, 79(1), 99.

Harloe, M., & Perry, B. (2004) Universities, localities and regional development: The emergence of the 'Mode 2' university? International Journal of Urban and Regional Research, 28(1), 212.

Harrison, B. (1972) Education, training, and the urban ghetto. Baltimore: Johns Hopkins University Press.

Harrison, B. (1994a) Lean and mean: The changing landscape of corporate power in the age of flexibility. New York: Basic Books.

Harrison, B. (1994b) The small firms myth. California Management Review, 26(3), 142–159.

Harvey, D. (1989) The condition of postmodernity: An enquiry into the origins of cultural change. Oxford, UK ; Cambridge, Mass.: B. Blackwell.

Harvey, D. (1990) Between space and time: Reflections on the geographical imagination. Association of American Geographers. Annals of the Association of American Geographers, 80(3), 418.

Heenan, D. A., & Perlmutter, H. V. (1979) Multinational organizational development: A social architectural approach. Reading, Mass.: Addison-Wesley.

Hendry, C., Brown, J., & Defillippi, R. (2000) Regional clustering of high technology-based firms: Opto-electronics in three countries. Regional Studies, 34(2), 129.

Herod, A. (1997) From a geography of labor to a labor geography: Labor's spatial fix and the geography of capitalism. Antipode, 29(1), 1–31.

Hicks, D., & Hegde, D. (2005) Highly innovative small firms in the markets for technology. Research Policy, 34(5), 703.

Hill, E. W., & Brennan, J. F. (2000) A methodology for identifying the drivers of industrial clusters: The foundation of regional competitive advantage. Economic Development Quarterly, 14(1), 65.

Hill, L. (2004) Can media artists survive media consolidation? The Journal of the Caucus of Television Producers, Writers and Directors, XXII(Summer), 17–21.

Holland, S. (1976) Capital versus the region. London: Macmillan.

Holt, J. (2001) In deregulation we trust: The synergy of politics and industry in Reagan-era Hollywood. Film Quarterly, 55(2), 22–29.

Hudson, R. (2001) Producing places. New York: Guilford Press.

Hudson, R., & Williams, A. M. (1999) Divided Europe: Society and territory. London; Thousand Oaks, California.: SAGE Publication.

Immergluck, D. (1998) Neighborhood economic development and local working: The effect of nearby jobs on where residents work. Economic Geography, 74(2), 170.

Ionescu-Heroiu, M. (2007) From "please come" to "you're not welcome anymore": Offshoring of software and services to emerging markets. Unpublished Work. Cornell University.

Isard, W. (1956) Location and space-economy; a general theory relating to industrial location, market areas, land use, trade, and urban structure. Cambridge: Technology Press of Massachusetts Institute of Technology Wiley New York.

Jacobs, S. (2002, July 19) What's next for optics? Rochester Business Journal, pp. 20–21.

Jacoby, S. M. (1997) Modern manors: Welfare capitalism since the New Deal. Princeton, N.J.: Princeton University Press.

Jacoby, S. M. (2005) The embedded corporation: Corporate governance and employment relations in Japan and the United States Princeton, N.J.: Princeton University Press.

Jonas, A. E. (1996) Local labour control regimes: Uneven development and the social regulation of production. Regional Studies, 30(4), 323.

Jones, M. (2002) Motion picture production in California. Sacramento, Calif.: California Research Bureau.

Kapstein, E. B. (1992) The political economy of national security: A global perspective. New York: McGraw-Hill.

Katz, H. C., & Darbishire, O. (1999) Converging divergences: Worldwide changes in employment systems. Ithaca, N.Y.: Cornell University Press.

Kenney, M., & Patton, D. (2005) Entrepreneurial geographies: Support Networks in three high-technology industries. Economic Geography, 81(2), 201.

Kipnis, S., & Huffstutler, C. (1990) Productivity trends in the photographic equipment and supplies industry. Monthly Labor Review, Bureau of Labor Statistics, 113(6), 39–48.

Kogut, B. (1984) Normative observations on the international value-added chain and strategic groups. Journal of International Business Studies, 15(2), 151–167.

Kogut, B. (1985) Designing global strategies: Comparative and competitive value-added chains. Sloan Management Review, 26(4), 15–28.

Krugman, P. R. (1991) Geography and trade. Cambridge, Mass.: MIT Press.

Krumholz, N., & Clavel, P. (1994) Reinventing cities: Equity planners tell their stories. Philadelphia, Penn.: Temple University Press.

Lagendijk, A. & Päivi Oinas (2005) Proximity, distance and diversity: issues on economic interaction and local development. Aldershot, England; Burlington, Vt.: Ashgate.

Lane, Christel (1995) Industry and society in Europe: stability and change in Britain, Germany, and France. Cheltenham: Edward Elgar.

Lash, S., & Urry, J. (1987) The end of organized capitalism. Madison, Wis.: University of Wisconsin Press.

Latour, B. (1993) Have we ever been modern? Hemel Hempstead, U.K.: Harvester Wheatsheaf.

Lautsch, B. A. (2003) The influence of regular work systems on compensation for contingent workers. Industrial Relations, 42(4), 565.

Lazonick, W., & O'Sullivan, M. (1997) Investment in innovation: Corporate governance and employment: is prosperity sustainable in the United States? Annandale-on-Hudson, N.Y.: Bard Publications Office.

Lee, K. (2002, March 19–23) Remaking the geography of southern California's film and television industry. Paper presented at the Annual Meeting of the American Association of Geographers, Los Angeles.

Lorenzen, M., & Mahnke, V. (2002) Global strategy and the acquisition of local knowledge: How MNCs enter regional knowledge clusters. Paper presented at the DRUID Summer Conference on Industrial Dynamics and the New and Old Economy – Who is Embracing Whom?, Copenhagen.

Lovering, J. (1999) Theory led by policy: The inadequacies of the 'New Regionalism' (illustrated from the case of Wales) International Journal of Urban and Regional Research, 23(2), 379.

Lovering, J. (2001) The coming regional crisis (and how to avoid it) Regional Studies, 35(4), 349–354.

Malecki, E. J. (2005) The United States: Still on top? European Planning Studies, 13(8), 1173.

Malmberg, A., & Power, D. (2005) On the role of global demand in local innovation processes. In G. Fuchs & Shapira, P. (eds), Rethinking Regional Innovation and Change; Path Dependency or Regional Breakthrough. New York: Springer.

Malmberg, A., Solvell, O., & Zander, I. (1996) Spatial clustering, local accumulation of knowledge and firm competitiveness. Geographiska Annaler. Series B, 78(2), 85–97.

Maltby, R. (1998) 'Nobody knows everything': Post-classical historiographies and consolidated entertainment. In S. Neale & M. Smith (eds), Contemporary Hollywood cinema (pp. 21–44) New York: Routledge.

Manly, L. (2005, June 20) Networks and the outside producer: Can they co-exist? New York Times.

Manpower Inc. (2003) Manpower employment outlook survey. Retrieved January 2003, from http://www.manpower.com/mpcom/content.jsp?articleid=64

Markusen, A. R. (1985) Profit cycles, oligopoly, and regional development. Cambridge, Mass.: MIT Press.

Markusen, A. R. (1987) Regions: The economics and politics of territory. Totowa, N.J.: Rowman & Littlefield.

Markusen, A. R. (1991) The rise of the Gunbelt: The military remapping of industrial America. New York: Oxford University Press.

Markusen, A. R. (1996) Sticky places in slippery space: A typology of industrial districts. Economic Geography, 72(3), 293.

Markusen, A. R. (1999) Fuzzy concepts, scanty evidence, policy distance: The case for rigour and policy relevance in critical regional studies. Regional Studies, 33(9), 869.

Markusen, A. R. (2001a) The activist intellectual. Antipode, 33(1), 39–49.

Markusen, A. R. (2001b) Targeting occupations in regional and community economic development. 2003, http://www.hhh.umn.edu/projects/prie/working.htm

Markusen, A. R. (2004) Targeting occupations in regional and community economic development. American Planning Association. Journal of the American Planning Association, 70(3), 253.

Marshall, A. (1920) Principles of economics, an introductory volume (8th edn) London: Macmillan.

Martin, R. (2001) Geography and public policy: The case of the missing agenda. Progress in Human Geography, 25(2), 189–210.

Martin, R., & Sunley, P. (1997) The post-Keynsian state and the space economy. In R. Lee & J. Wills (eds), Geographies of Economies (pp. 278–289) London: Arnold.

Martin, R., & Sunley, P. (1998) Slow convergence? Economic Geography, 74(3), 201–227.

Martin, R., & Sunley, P. (2001) Rethinking the 'economic' in economic geography: Broadening our vision of losing our focus. Antipode, 33(2), 148–161.

Martin, R., & Sunley, P. (2006) Path dependence and regional economic evolution. *Journal of Economic Geography*, 6(4), 395.

Massey, D. (1979) In what sense a regional problem? Regional Studies, 13(2), 233–243.

Massey, D. (1984) Spatial divisions of labour: Social structures and the geography of production. London: Macmillan.

Massey, D. (2000) Practising policy relevance. Transactions of the Institute of British Geographers, 24, 131–134.

Massey, D., & Meegan, R. (1982) The anatomy of job loss: The how, why, and where of employment decline. London: Methuen.

Mattera, P., Purinton, A., McCourt, J., Hoffer, D., Greenwood, S., & Talanker, A. (2004) Shopping for subsidies: How Wal-Mart uses taxpayer money to finance its never-ending growth: Good Jobs First.

McDowell, L. (1997) A tale of two cities? Embedded organizations and embodied workers in the City of London. In R. Lee & J. Wills (eds), Geographies of Economies. London: Arnold.

McKelvey, B. (1956) Rochester: The quest for quality, 1890–1925. Cambridge: Harvard University Press.

McKelvey, B. (1973) Rochester on the Genesee: The growth of a city (1st edn) Syracuse, N.Y.: Syracuse University Press.

McLean, L. (2001) Rational choice and British politics: An analysis of rhetoric and manipulation from Peel to Blair. Oxford: Oxford Univeristy Press.

Miller, T., Govil, N., McMurria, J., & Maxwell, R. (2001) Global Hollywood. London: British Film Institute.

Monitor. (1999) US runaway film and television production study report, commissioned by the directors Guild of America and Screen Actors Guild. Santa Monica, Calif.: Monitor Company.

Montpool, T. (1998) Casting labour: The dynamics of Toronto's film and television production industry. Thesis. University of Waterloo, Waterloo, Ontario, Canada.

Morgan, K. (1997) The learning region: Institutions, innovation and regional renewal. Regional Studies, 31(5), 491.

Morgan, K. (2004) The exaggerated death of geography: Learning, proximity and territorial innovation systems. Journal of Economic Geography, 4(1), 3.

Motion Picture Association of America. (2005) US entertainment industry market statistic report. Los Angeles, Calif.: Motion Picture Association of America.

Mudambi, R., & Helper, S. (1998) The 'close but adversarial' model of supplier relations in the US auto industry. Strategic Management Journal (1986–1998), 19(8), 775.

National Association of Manufacturers (2005) 2005 Skills gap report a survey of the American manufacturing workforce. Washington, DC: National Association of Manufacturers.

National Science Board (2004) U.S. and International research and development: Funds and technology linkages: International R&D trends and comparisons. Washington: National Science Board. http://www.nsf.gov/statistics/seind04/c4/c4s4.htm (downloaded May 21, 2007).

Nelson, R., & Winter, S. (1982) An evolutionary theory of economic change. Cambridge, Mass.: Harvard University.

New York State Department of State Division of State Planning, New York State Department of Commerce, & New York State Department of Labor (2000) Industrial directory of New York State; directory of manufacturing and mining firms in New York State. Albany, N.Y.: State of New York Department of Commerce.

Offe, C. (1984) Contradictions of the welfare state. Cambridge, Mass.: MIT Press.

Ohmae, K. (1995) The end of the Nation State – the rise of regional economies. London: HarperCollins.

Orfield, M. (1997) Metropolitics: A regional agenda for community and stability. Washington, DC: The Brookings Institution Press.

Osterman, P. (1999) Securing prosperity: The American labor market: How it has changed and what to do about it. Princeton, N.J.: Princeton University Press.

Ottaviano, G. I., & Puga, D. (1998) Agglomeration in the global economy: A survey of the "new economic geography". The World Economy, 21(6), 707.

Patel, P., & Pavitt, K. (1997) The technological competencies of the world's largest firms: Complex and path-dependent, but not much variety. Research Policy, 26(2), 141.

Pauly, L. W., & Reich, S. (1997) National structures and multinational corporate behavior: Enduring differences in the age of globalization. International Organization, 51(1), 1.

Pearce, D. W. (ed.) (1992) The MIT Dictionary of Modern Economics. Cambridge, Mass.: The MIT Press.

Peck, J. (1992) Labor and agglomeration: Control and flexibility in local labor markets. Economic Geography, 68(4), 325.

Peck, J. (1996) Work-place: The social regulation of labor markets. New York: Guilford Press.

Peck, J., & Theodore, N. (2001) Contingent Chicago: Restructuring the spaces of temporary labor. International Journal of Urban and Regional Research, 25(3), 471.

Pendall, R. (2000) Local land use regulation and the chain of exclusion. American Planning Association. Journal of the American Planning Association, 66(2), 125.

Pendall, R., & Christopherson, S. (2004) Losing ground: Income and poverty in Upstate New York, 1980–2000. Washington, DC: The Brookings Institution.

Pendall, R., Drennan, M. P., & Christopherson, S. (2004) Transition and renewal: The emergence of a diverse Upstate economy. Washington, DC: The Brookings Institution.

Penrose, E. T. (1995) The theory of the growth of the firm (3rd edn) Oxford; New York: Oxford University Press.

Peterson, G., & Vroman, W. (eds) (1992) Urban labor markets and job opportunity. Washington, DC: The Urban Institute Press.

Pethel, B. (2002) Canada's movie, TV subsidies under fire. Toronto Star p. A23.

Phelps, E. S. (1997) Rewarding work: How to restore participation and self-support to free enterprise. Cambridge, Mass.: Harvard University Press.

Pike, A. (2005) "Shareholder value" versus the regions: The closure of the Vaux Brewery in Sunderland. Journal of Economic Geography, 6(2), 201–222.

Piore, M. J. (1990) in Sengenberger, W., Loveman, G., Piore, M. J., & International Institute for Labour Studies. The re-emergence of small enterprises: industrial restructuring in industrialised countries. Geneva: International Institute for Labour Studies.

Piore, M. J., & Sabel, C. (1983) Italian small business development: Lessons for U.S. industrial policy. In J. Zysman & L. Tyson (eds), American industry in international competition: Government policies and corporate strategies. Ithaca, N.Y.: Cornell University Press.

Piore, M. J., & Sabel, C. F. (1984) The second industrial divide: Possibilities for prosperity. New York: Basic Books.

Pollard, J., Henry, N., Bryson, J., & Daniels, P. (2000) Shades of grey: Geographers and policy. Transactions of the Institute of British Geographers, 24, 243–248.

Pollard, J., & Storper, M. (1996) A tale of twelve cities: Metropolitan employment change in dynamic industries in the 1980s. Economic Geography, 72(1), 1.

Porter, M. E. (1990) The competitive advantage of nations. New York: Free Press.

Porter, M. E. (1990) The competitive advantage of nations. Harvard Business Review, 68(2), 73.

Porter, M. E. (1998) Competitive strategy: Techniques for analyzing industries and competitors: With a new introduction (1st Free Press edn) New York: Free Press.

Porter, M. E. (2000) Location, competition, and economic development: Local clusters in a global economy. Economic Development Quarterly, 14(1), 15.

Porter, M. E., & Stern, S. (2001) Innovation: Location matters. MIT Sloan Management Review, 42(4), 28.

Pred, A. (1977) City Systems in Advanced Economies. Berkeley, Calif.: University of California Press.

Prince, S. (2000) A new pot of gold: Hollywood under the electronic rainbow, 1980–1989. New York: Charles Scribner's Sons.

Pulido, L. (2000) Rethinking environmental racism: White privilege and urban development in Southern California. Association of American Geographers. Annals of the Association of American Geographers, 90(1), 12.

Quintas, P., Wield, D., & Massey, D. (1992) Academic-industry links and innovation: Questioning the science park model. Technovation, 12(3), 161.

Rast, J. (2006) Environmental Justice and the New Regionalism. Journal of Planning Education and Research, 25(3), 249.

Regini, M. (2000) Between Deregulation and Social Pacts: The Responses of European Economies to Globalization. Politics and Society, 28(1), 5–34.

Roe, M. J. (2003) Political determinants of corporate governance: Political Context, Corporate Impact. Oxford: Oxford University Press.

Rutherford, T., & Holmes, J. (2006) Entrepreneurship, knowledge, and learning in the formation and evolution of industrial clusters: The case of the Windsor, Ontario tool, die, and mould cluster. Journal of Entrepreneurship and Regional Development.

Saxenian, A. (1989) The Cheshire cat's grin: Innovation, regional development, and the Cambridge case. Berkeley, Calif.: Institute of Urban and Regional Development University of California at Berkeley.

Saxenian, A. L. (1994) Regional advantage: Culture and competition in Silicon Valley and Route 128. Cambridge, Mass.: Harvard University Press.

Schoenberger, E. J. (1997) The cultural crisis of the firm. Cambridge, Mass.: Blackwell Publishers.

Schoenberger, E. J. (1999) The firm in the region and the region in the firm. In T. Barnes & M. Gertler (eds), The New Industrial Geography: Regions, Regulation, and Institutions. New York: Routledge.

Schutz, E. A. (2001) Markets and power: the 21st century command economy. New York: M.E. Sharpe.

Scott, A. (1988a) Flexible production systems and regional development: The rise of new industrial spaces in North America and Western Europe. International Journal of Urban and Regional Research, 12, 171–186.

Scott, A. (1988b) Metropolis: From the division of labor to urban form. Berkeley, Calif.: University of California Press.

Scott, A. (1988c) New industrial spaces: Flexible production organization and regional development in North America and Western Europe. London: Pion.

Scott, A. (1992) The role of large producers in industrial districts: A case study of high technology systems houses in Southern California. Regional Studies, 26(3), 265–275.

Scott, A. (1998) Regions and the world economy: The coming shape of global production, competition, and political order. Oxford; New York: Oxford University Press.

Scott, A. (2005) On Hollywood: The place, the industry. Princeton: Princeton University Press.

Scott, A., & Storper, M. (2003) Regions, globalization, development. Regional Studies, 37(6 & 7), 579–593.

Sengenberger, W., Loveman, G., Piore, M. J., & International Institute for Labour Studies. (1990) The re-emergence of small enterprises: Industrial restructuring in industrialised countries. Geneva: International Institute for Labour Studies.

Sethi, S. P. (1970) Business corporations and the black man; an analysis of social conflict: The Kodak-FIGHT controversy. Scranton, Penn.: Chandler Pub. Co.

Sheppard, E. (2002) The spaces and times of globalization: Places, scale, networks, and positionalility. Economic Geography, 78(3), 307–330.

Soja, E. W. (1989) Postmodern geographies: The reassertion of space in critical theory. London: Verso.

Sokol, M. (2003) Regional dimensions of the knowledge economy: Implications for the New Europe. Doctoral dissertation, Newcastle Upon Tyne: Centre for Urban and Regional Development Studies, University of Newcastle Upon Tyne, England

Sternberg, E. (1992) Photonic technology and industrial policy. Albany, N.Y.: State University of New York Press.

Stone, K. (1973) The origins of job structures in the steel industry. In R. Edwards, M. Reich & D. Gordon (eds), Labor market segmentation. Toronto: D.C. Heath and Company.

Stone, K. (1981) The post-war paradigm in American labor law. The Yale Law Journal, 90(7), 1509.

Stone, K. (2001) The new psychological contract: Implications of the changing workplace for labor and employment law. UCLA Law Review, 48, 540–549.

Stone, K. (2004) From widgets to digits: Employment regulation for the changing workplace. New York: Cambridge University Press.

Storper, M. & Christopherson, S. (1985) The changing organization and location of the motion picture industry. Research Report R854, Graduate School of Architecture and Urban Planning, University of California, Los Angeles.

Storper, M. (1997) The regional world: Territorial development in a global economy. New York: Guilford Press.

Storper, M. (1999) The resurgence of regional economies. In T. Barnes & M. Gertler (eds), The new industrial geography: Regions, regulation, and institutions. London: Routledge.

Storper, M. (2001) The poverty of radical theory today: From the false promises of Marxism to the mirage of the cultural turn. International Journal of Urban and Regional Research, 25(1), 155.

Storper, M. (2002) Institutions of the learning economy. In M. Gertler & D. Wolfe (eds), Innovation and social learning. New York: Palgrave Mcmillan.

Storper, M., & Harrison, B. (1991) Flexibility, hierarchy and regional development: The changing structure of industrial production systems and their forms of governance in the 1990s. Research Policy, 20(5), 407.

Storper, M., & Scott, A. (1990) Work organisation and local labour markets in an era of flexible production. International Labour Review, 129, 573–591.

Swanstrom, T. (2001) What we argue about when we argue about regionalism. Journal of Urban Affairs, 23(5), 479.

Taylor, M., & Asheim, B. (2001) The concept of the firm in economic geography. Economic Geography, 77(4), 315.

Thrift, N. (1994) On the social and cultural determinants of international financial centres: The case of the City of London. In S. Corbridge, R. Martin & N. Thrift (eds), Money, power and space. Cambridge, Mass.: Blackwell.

Tiebout, C. M. (1962) The community economic base study. New York: Committee for Economic Development.

U.S. Bureau of Labor Statistics (2003) Union members summary. Washington, DC: Division of Labor Force Statistics, U.S. Department of Labor.

U.S. Bureau of Labor Statistics Data (1997–2003) Covered employment and wages (SIC, NAICS) Retrieved March 2003, September 2004, from http://data.bls.gov/labjava/outside.jsp?survey=ew, http://www.bls.gov/cew/, http://data.bls.gov/labjava/outside.jsp?survey=en

U.S. Department of Commerce (2001) Report on runaway production. Washington, DC: U.S. Department of Commerce.

Van Jaarsveld, D. D. (2000) Nascent organizing initiatives among high-skilled contingent workers: The Microsoft-WashTech/CWA case. Cornell University, Ithaca.

Van Jaarsveld, D. D. (2002) Changing work relationships in industrialized economics. Industrial & Labor Relations Review, 55(3), 557.

Van Jaarsveld, D. D. (2004) Collective representation among high-tech workers at Microsoft and beyond: Lessons from Wash Tech/CWA. Industrial Relations, 43(2), 364.

Vogel, R. K. (2002) Metropolitan planning organizations and the new regionalism: The case of Louisville. Publius, 32(1), 107.

Wadhwani, R. D. G. (1997) Kodak, FIGHT, and the definition of civil rights in Rochester, New York: 1966–1967. Historian: A Journal of History, 60(1), 59–75.

Walkowitz, D. J. (1978) Worker city, company town: Iron and cotton-worker protest in Troy and Cohoes, New York, 1855–84. Urbana, Ill.: University of Illinois Press.

Webber, M. J. (1982) Agglomeration and the regional question. Antipode, 14(2), 1–11.

Weber, Rachel (1998) The state as corporate stakeholder: governing the decline of the military-industrial complex. Ph.D. dissertation, Cornell University.

Webster, R., A. (2005, 9 May) Proposed cap on tax credits would be harmful for LA's film industry. New Orleans City Business

Weinstein, B., & Clower, T. (2000) Filmed entertainment and local economic development: Texas as a case study. Economic Development Quarterly 14(4), 795–819.

Weir, M., Wolman, H., & Swanstrom, T. (2005) The calculus of coalitions: cities, suburbs, and the metropolitan agenda. Urban Affairs Review, 40(6), 730.

Wernerfelt, Birger (1984) A resource-based view of the firm. Summary. Strategic Management Journal, 5(2), 171.

West, G. P., & DeCastro, J. (2001) The achilles heel of firm strategy: Resource weaknesses and distinctive inadequacies. The Journal of Management Studies, 38(3), 417.

Whitley, R. (1992) European business systems: firms and markets in their national contexts. London: Sage.

Whitley, R. (1999) Divergent capitalisms: the social structuring and change of business systems. Oxford: Oxford University Press.

Winseck, D. (2002) Netscapes of power: Convergence, consolidation and power in the Canadian mediascape. Media, Culture and Society, 24: 795–819.

Wolfe, D. (1999) Harnessing the region: Changing perspectives on innovation policy in Ontario. In T. Barnes & M. Gertler (eds), New Industrial Geography; Regions, Regulation, and Institutions. New York: Routledge.

Wolfe, D., & Holbrook, J. A. (2000) Innovation, institutions and territory: Regional innovation systems in Canada. Montreal; Ithaca, N.Y.: Published for the School of Policy Studies Queen's University by McGill-Queen's University Press.

Wolf-Powers, L. (2001a) Generating jobs: How to increase demand for less-skilled workers. American Planning Association. Journal of the American Planning Association, 67(2), 235.

Wolf-Powers, L. (2001b) Information technology and urban labor markets in the United States. International Journal of Urban and Regional Research, 25(2), 427.

Yeung, H. W. (2002) Industrial geography: Industrial restructuring and labour markets. Progress in Human Geography, 26(3), 367–379.

Zook, M. & S. Brunn. (2006) From podes to antipodes: positionalities and global airline geographies, Annals of the Association of American Geographers, 96(3), 471–490.

Index

Note: *italic* page numbers denote references to figures/tables.